ISBN 978-0-265-98281-5
PIBN 10918973

SCHOOL ARITHMETICS

PRIMARY BOOK

BY

FLORIAN CAJORI

New York
THE MACMILLAN COMPANY
1914

Set up and electrotyped. Published June, 1914.

Norwood Press
J. S. Cushing Co. — Berwick & Smith Co.
Norwood, Mass., U.S.A.

PREFACE

THIS Primary arithmetic presents the subject in a way that is attractive to a pupil. Various numerical plays, games, and drill devices are introduced to encourage the pupil's self-activity. This is accomplished without sacrifice of serious intent. Frequent reviews aid the pupil in retaining the facts and processes he has discovered.

The technique of arithmetic is simplified in several cases, with the view of securing greater economy of effort. For instance, the quotient which in other works is written below the dividend in short division, and above the dividend in long division, is here always written above.

The practice in this book is arranged so as to establish the pupil in habits of accuracy and to develop reasonable speed in the correct performance of the operations. The exercises and problems are progressive tests. Special tests for the measuring of the arithmetical abilities of the children are given at the close of the book.

For coöperation in the preparation of this and the other books of the series, the author is indebted to several teachers in the public schools of Colorado Springs, particularly to Mrs. L. D. Coffin, Mrs. I. J. Lewis, Miss Minnie L. McCall, and Miss Edna Kinder.

FLORIAN CAJORI.

CONTENTS

PART ONE

PART TWO

PART THREE

PRIMARY BOOK

PART ONE

INTRODUCTION

Exercise in Counting

1. 1. Point to one block. Two blocks. Three blocks. Four blocks. Five blocks.

2. Count six desks. Count seven pupils. Count the letters in the word *children*, in the word *exercises*.

3. Count to ten.

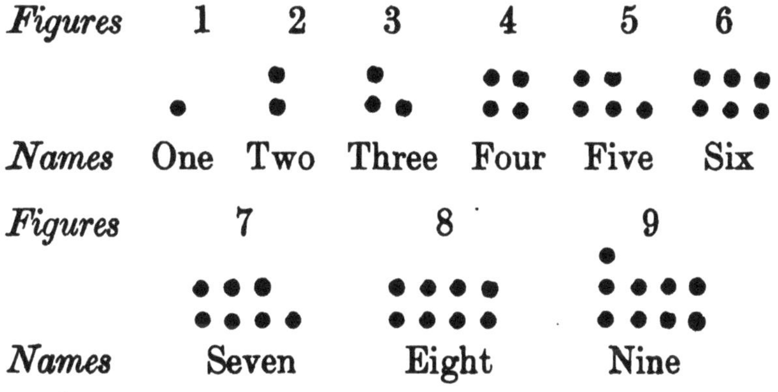

4. Show by splints how many things in 1. In 2. In 3. In 4. In 5. In 6. In 7. In 8. In 9.

5. Write the names of each of the nine figures.

ADDITION AND SUBTRACTION

Oral Problems

2. 1. How many balls are

2.  and  are how many balls?

3. How many apples are

4. How many marbles are 6 marbles and 3 marbles?

5. How many cents are 2 cents and 3 cents?

6. How many cherries are 5 cherries and 2 cherries?

7. John has 3 apples and gives away 2. How many apples has he left?

8. Mary has 5 pears and gives 1 to Emma. How many pears has Mary left?

9. How many more books are 4 books than 2 books?

10. James had 5 marbles and lost 1. How many marbles has he left?

11. Make five problems. Give the answers.

Exercise—Oral

3. Add the numbers:

1.	2.	3.	4.	5.
1	2	4	5	3
1	1	1	1	1

6.	7.	8.	9.	10.
7	1	8	6	6
1	2	1	1	2

11.	12.	13.	14.	15.
4	2	5	2	1
2	2	2	5	2

16.	17.	18.	19.	20.
3	2	4	2	7
2	3	4	4	2

21.	22.	23.	24.	25.
6	2	3	4	3
3	7	3	3	4

26.	27.	28.	29.	30.
5	6	3	2	5
4	3	6	5	3

31.	32.	33.	34.	35.
1	1	1	3	1
5	8	7	4	6

Oral Exercises

4. Take away the lower figure from the upper:

1.	2.	3.	4.	5.	6.	7.
2	3	5	8	7	4	6
1	1	1	1	1	1	1

8.	9.	10.	11.	12.	13.	14.
9	6	3	5	4	7	8
1	2	2	2	2	2	2

Written Exercises

5. Add the numbers in each column:

1.	2.	3.	4.	5.	6.	7.
1	1	1	2	2	2	1
1	2	3	1	2	3	4
1	1	1	2	2	2	1

8.	9.	10.	11.	12.	13.	14.
7	5	3	3	5	2	2
1	1	4	3	1	2	5
1	2	1	2	1	1	1

Written Exercises

6. Take away the lower figure from the upper:

1.	2.	3.	4.	5.	6.	7.
4	5	7	8	6	4	8
3	3	3	3	3	2	2

8.	9.	10.	11.	12.	13.	14.
9	2	3	5	6	8	9
3	2	3	4	4	4	4

15.	16.	17.	18.	19.	20.	21.
9	6	7	9	8	9	9
3	5	5	5	6	6	0

22.	23.	24.	25.	26.	27.	28.
7	9	6	6	8	8	8
6	2	6	2	7	0	8

29.	30.	31.	32.	33.	34.	35.
9	9	1	5	8	7	7
7	9	0	5	5	4	22

NOTE.—If subtraction has not been previously taught, the *Austrian Method* may be used. This method consists in finding the number that must be added to the subtrahend in order to make the minuend. According to this method, 15 – 8 is solved as follows:

8 and 7 make 15. Write the 7 in the column of ones.

If subtraction has been previously taught by the older method, that method should be continued, thus:

8 from 15 are 7. Write the 7 in the column of ones.

Comparisons — Oral

7. 1. Line *A* is 1 inch long. How long is *B*? How long is *C*?

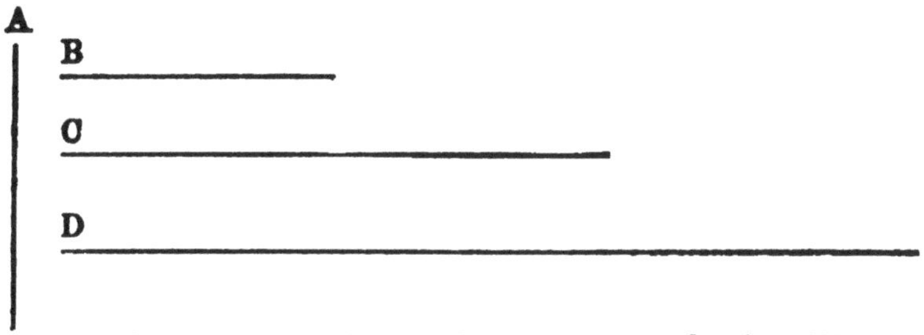

A

B

C

D

2. How many lines the length of *A* will make a line the length of *C*?

3. Are two lines the length of *B* more than *D* or less?

4. Are *A* and *B* together equal to *C*?

5. Are *A* and *C* together more than *D* or less?

READING AND WRITING NUMBERS TO 20

8. 1. How many splints are there in

Ten is written 10.

2. How many splints in

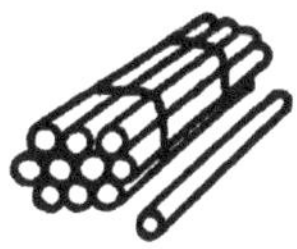

3. How many splints are *ten* splints and *one* splint? *Eleven* is written 11.

4. Which of the 1's in 11 stands for 1 splint?

5. How many splints in *Twelve* is written 12.

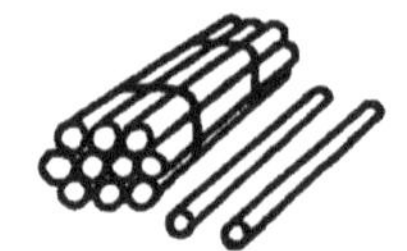

6. In 12, which bundle does the figure to the right stand for?

7. What does the figure to the left stand for?

8. How many splints in *Thirteen* is written 13. What does the 3 stand for? The 1?

9. How shall I write the number that stands for *fourteen* splints? For how many splints does each figure stand?

10. In 15, which figure stands for *ones?* Which for *ten?*

11. In 16, what does the 6 stand for? The 1?

12. What does each figure stand for in 17? In 18? In 19?

The figure 0 is called *zero* or *naught*. It stands for *nothing* or *not any*.

Written Exercises

9. 1. Write in words all numbers from *ten* to *nineteen*.

2. Write in figures all numbers from *ten* to *nineteen*.

3. Write in figures the number of letters in the following words:

arithmetic	arithmetical	horse
arithmetics	brother	geographies

4. Write in figures the number of dots in the first group; the number of dots in the second group.

● ● ● ● ● ● ● ● ● ● ● ●
● ● ● ● ● ● ● ● ● ● ● ●
● ● ● ● ● ● ● ● ● ● ● ●

Addition — Oral Problems

10. 1. John has 4 apples and Joseph has 2. How many apples have both together?

2. Mary had five apples. Lucy gave her four more. How many apples had Mary then?

3 A girl found five eggs in one nest and two eggs in another nest. How many eggs did she find in all?

4. James has three cents His father gives him five more. How many has James now?

5. George puts two books on the table. Charles puts five. How many books are now on the table?

6. How many balls are six balls and three balls?

7. One word is spelled with seven letters, another with four. How many letters in both words?

8. John found nine shells on the beach. Mary found two. How many shells did they both find?

Oral Exercise

11. Add 1 to each of the following:

3	6	9	8	5	7	10	11
15	13	2	12	14	17	4	16

Add 10 to each of the following:

5 6 1 4 3 2 7 8 9

Give answers quickly:

1.	**2.**	**3.**	**4.**	**5.**	**6.**
1 +2	2 +2	4 +1	4 +0	5 +1	2 +1
7. 2 +0	**8.** 1 +4	**9.** 0 +4	**10.** 1 +5	**11.** 3 +1	**12.** 3 +2
13. 2 +4	**14.** 4 +3	**15.** 2 +5	**16.** 2 +3	**17.** 4 +2	**18.** 3 +4

3
+ 2 means 3 and 2, or 3 *plus* 2. The sign + is called *plus*. The answer is called the *sum*.

Oral Exercises

12. Answer by lines and then by columns:

1.	**2.**	**3.**	**4.**	**5.**	**6.**
5 +4	6 +1	6 +4	7 +2	3 +7	0 +5
7. 4 +5	**8.** 1 +6	**9.** 4 +6	**10.** 2 +7	**11.** 7 +3	**12.** 5 +3
13. 5 +5	**14.** 2 +6	**15.** 6 +3	**16.** 7 +1	**17.** 7 +4	**18.** 3 +5

Oral Problems

13. 1. Sarah picked 8 apples from one branch of a tree, and 2 apples from another. How many apples did she pick?

2. A farmer has 3 horses in one field and 8 horses in another field. How many horses has he in both fields?

3. In one automobile 4 persons may ride; in another, 8 persons. How many persons may ride in both automobiles?

4. Martha placed 6 books in one strap and 8 books in another. How many books has she in both straps?

5. How many are 8 cents and 7 cents?

6. Lucy found 6 eggs in one nest and 7 eggs in another. How many eggs did Lucy find?

7. There are 7 boys and 8 girls in one class. How many pupils in the class?

8. How many cents will it take to buy a 5-cent ball and a 9-cent kite?

9. Ned has 15 small marbles and 5 large ones. How many marbles has he?

10. Mary picked 7 violets and Martha 12. How many did the two girls pick?

11. A boy buys a 5-cent note book, a 2-cent pencil, and a 3-cent ruler. How much does he pay for all?

Oral Exercises

14. Add at sight:

1.	2.	3.	4.	5.	6.	7.
12	7	6	11	4	8	9
+2	+2	+2	+4	+4	+1	+0

8.	9.	10.	11.	12.	13.	14.
4	7	9	8	5	6	10
+6	+5	+4	+3	+9	+7	+9

15.	16.	17.	18.	19.	20.	21.
10	11	13	16	12	14	15
+7	+5	+2	+3	+7	+4	+4

Oral Problems

15. 1. $5 + $4 = $——; $10 + $8 = $——; $8 + $11 = $——.

2. 5 chairs + 2 chairs = —— chairs; 10 pears + 6 pears = —— pears.

3. 12 men + 5 men = —— men; 11 tacks + 4 tacks = —— tacks.

4. Make problems about 3 + 7 = ——; 7 + 4 = ——.

5. Make a problem about 2 + 3 + 6 = ——. 7¢ + 3¢ + 5¢ = ——¢.

The dollar sign is $. $7 + $4 = $11. It is read 7 dollars *and* 4 dollars *are* 11 dollars. The cent sign is ¢.

Written Exercises

16. Copy and add:

1.	2.	3.	4.	5.	6.	7.
3	4	5	7	8	11	0
5	9	1	6	0	6	9
6	0	3	2	4	0	1
1	3	9	1	5	1	7

Add each column beginning at the bottom. Do not name the numbers added, but say only the partial sums. Thus, in the first example, 7, 12, 15.

To make sure that you have made no mistake, add from the top down. If the result is the same as the one first obtained, the work should be correct.

8. Make six examples like those in this exercise.

Measures of Length — Exercises

17. 1. Without using a ruler, draw on paper a straight line 2 in. long.

2. Draw a straight line 4 in. long.

3. Draw a straight line 3 in. long.

4. Measure the lines with a foot rule. How long are they?

5. Draw on the blackboard a straight line 1 ft. long.

6. Draw on the blackboard a straight line 2 ft. long.

7. Measure these lines with a foot rule. How long are they?

8. Use a foot rule and a yardstick to measure other lines in the room.

MEASURES OF LENGTH	
12 inches (in.)	= 1 foot (ft.)
3 feet	= 1 yard (yd.)

NOTE. — Cut strips from cardboard the length and width of a foot rule. Fold this strip in half and mark the inches on each half. Save this measure for future lessons in measurement.

Drill in Counting and Addition

18. Count:

1. Begin with 1 and count by 2's to 21. Thus: 1, 3, 5.
2. Begin with 2 and count by 2's to 20.
3. Begin with 1 and count by 3's to 19.
4. Begin with 2 and count by 3's to 20.
5. Count by 2's from 12 to 2. By 3's.
6. Read quickly: 11, 14, 17, 16, 19, 12, 15, 18.
7. Add:

5	6	7	8	7	6	5	5
1	2	3	4	5	5	8	9

8. Give the answers:

$5+2=?$	$9+4=?$	$10+6=?$
$9+3=?$	$4+9=?$	$9+2=?$
$8+5=?$	$6+7=?$	$7+3=?$

Subtraction — Oral Problems

19. 1. George had 6 cents and lost 2 of them. How many cents has he left?

2. A farmer had 7 horses and sold 3. How many has he left?

3. George has 9 chickens. John has 7. George has —— more than John?

4. A boy had 8 goldfish. Three died. How many lived?

5. Lucy is 10 years old. Mary is 4 years younger. How old is Mary?

6. Five blocks from 10 blocks leave —— blocks?

7. Carrie makes 12 dots on a slate. Albert erases 3 of them. How many are left?

8. Eleven dots are 3 more than —— dots?

9. There are 18 pupils in a class. Ten are boys. How many are girls?

Oral Exercises

20. Subtract the lower number from the upper:

1. 5	2. 7	3. 8	4. 6	5. 9	6. 10	7. 10
−4	−5	−3	−2	−7	−3	−4
8. 13	9. 13	10. 13	11. 13	12 13	13. 13	14. 13
−3	−4	−5	−6	−7	−8	−9
15. 14	16. 14	17. 14	18. 14	19. 14	20. 14	21. 14
−3	−7	−9	−6	−8	−5	−4

The exercises on page 13 are examples in *subtraction.*

$$\begin{array}{rl} 5 & \text{is the } \textit{minuend}, \\ -4 & \text{is the } \textit{subtrahend}, \\ \hline 1 & \text{is the } \textit{difference} \text{ or } \textit{remainder}. \end{array}$$

The sign − is called *minus.*

Oral Problems

21. 1. How many buttons are 8 buttons less 5 buttons?

2. Jane put 7 blocks on the table. Kate takes away 4. How many blocks are left?

3. Charles picked 12 peaches. He gave 5 to his sister. How many has he left?

4. How many more letters are there in the word *Washington* than in the word *Lincoln?*

5. Mary has 13 cents. Kate has 10 cents. How many more cents has Mary than Kate?

6. Sixteen apples are 6 apples and —— apples?

7. Fifteen lemons are 6 lemons and —— lemons?

8. A milkman has 12 quarts of milk in one can and 8 quarts in another can. How many more quarts are there in the first than in the second can?

9. A baseball team lost 12 games out of 21 games. How many did it win?

10. Frank had 15 cents in his bank. He drew out 10 cents. How much remains in the bank?

Subtraction — Exercise

22. Give the answers quickly:

1.	2.	3.	4.	5.	6.	7.
19	19	19	19	19	19	19
−9	−8	−6	−7	−4	−5	−3

8.	9.	10.	11.	12.	13.	14.
12	19	17	18	16	14	13
−3	−4	−5	−9	−6	−8	−4

15.	16.	17.	18.	19.	20.	21.
16	17	15	13	11	18	14
−7	−6	−6	−9	−4	−7	−4

Drill Exercise — Money

23. 1. 20¢ − 10¢ = ?¢; 19¢ − 8¢ = ?¢.

2. $17 − $5 = $?; $13 − $4 = $?

3. 8 cents *from* 10 cents = ? cents.

4. $11 *from* $19 = $?

5. 14¢ *less* 7¢ = ?¢; 19¢ − 7¢ = ?¢.

6. Use toy money to make additional problems.

7. Use the method of making change to find the answers.

A TABLE OF MONEY	
10 cents (¢)	= 1 dime
10 dimes	= 1 dollar ($)

NOTE. — Make toy money from cardboard. Draw and mark circles to represent a cent, a nickel, a dime, and a quarter. This toy money may be used in making and solving problems.

Oral Exercises

24. Find the missing numbers:

1. $7+?=8$ 2. $9+?=11$ 3. $10+?=13$

4.	5.	6.	7.	8.	9.	10.
14	15	17	18	15	9	19
−?	−?	−?	−?	−?	−?	−?
6	11	13	14	15	2	12

11.	12.	13.	14.	15.	16.	17.
13	19	7	13	14	19	19
−?	−?	−?	−?	−?	−?	−?
11	13	5	11	10	10	9

Oral Problems

25. 1. Fred has 13 white rabbits and 4 brown rabbits. How many rabbits has he in all?

2. Nine birds lighted on a roof. One flew away, and then 3 flew away. How many flew away? How many were left?

3. Robert had $ 4. His father gave him $1 more. He bought his mother a Christmas present for $ 2. How many dollars has he left?

4. Yesterday there were 3 roses on a bush. To-day 4 more bloomed. Margaret picked 2. How many are still on the bush?

5. Charles got 4 picture cards from his father and 7 from his mother. He gave away 2. How many has he now?

6. Mary passes 9 houses on her way to school and Margaret passes 4. How many more does Mary pass than Margaret?

7. How many books must I add to 8 books to make 17 books?

8. Harry caught 14 fishes and sold 5 of them. How many fishes has he left?

9. There are 9 trees in one row and 7 trees in another row. How many trees in both rows?

10. Arthur is 7 years old. William is 18 years old. How many years older is William than Arthur?

11. Charles paid 12¢ for paper and 6¢ for pencils. How much did he pay for paper and pencils? How much more did the paper cost than the pencils?

12. Ellen has 4 pins and needs 11. How many more must she get to make 11?

13. Alfred has 13 marbles. His brother has 6. How many have both? How many less than Alfred has his brother?

14. Richard had 6 peaches. His mother gave him 4 more. He ate 3. How many were left?

15. Walter had $5. His father gives him $4. How many dollars had he after spending $3?

16. Nettie has 5 roses and as many pansies. She loses six of the flowers. How many has she left?

Drill Device — The Ladder

26. 1. Begin at the bottom; add 2 to each number, as you go up the ladder.

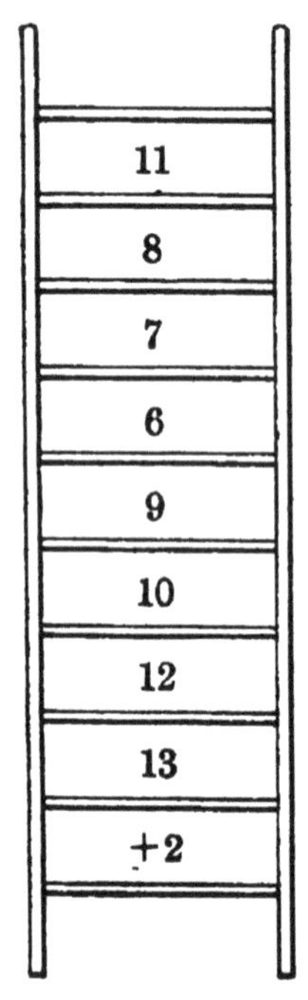

2. Begin at the top; subtract 2 from each number as you come down.

3. Add 3 as you go up; subtract 3 as you go down.

4. In the same way use 4 and 5.

Drill devices like the ladder may be drawn on cardboard or placed upon the blackboard and used for frequent drills. By changing the number to be added or subtracted, the use of the drill may be extended.

Try to reduce the time required in each drill.

This exercise may be used as a game. Each player begins at the bottom of the ladder and subtracts until he misses. The object of the game is to go up and come down without a mistake.

Review — Written Exercises

27. Write in figures:

1. Seven, nine, fourteen, sixteen, ten.
2. Nineteen, eighteen, five, six, fifteen.
3. Seventeen, thirteen, eleven, eight, twelve.

Written Exercises

28. Write in words:

1.	4	14	10	11	9	19	7	17
2.	8	18	6	16	5	15	3	13

Oral Exercises

29. Give the answers:

1.	7	**2.**	9	**3.**	6	**4.**	7	**5.**	7
	+7		+9		+6		−5		−3
6.	9	**7.**	9	**8.**	9	**9.**	7	**10.**	8
	−5		−3		+3		−6		−6
11.	9	**12.**	7	**13.**	6	**14.**	5	**15.**	3
	−6		+6		+7		+3		+5
16.	4	**17.**	8	**18.**	18	**19.**	7	**20.**	17
	+5		+1		+1		+2		+2
21.	9	**22.**	19	**23.**	8	**24.**	18	**25.**	6
	−7		−7		−5		−5		+2
26.	16	**27.**	7	**28.**	17	**29.**	5	**30.**	15
	+2		−5		−5		+3		+3
31.	6	**32.**	16	**33.**	5	**34.**	15	**35.**	4
	−4		−4		−3		−3		−4
36.	14	**37.**	6	**38.**	16	**39.**	4	**40.**	14
	−4		−2		−2		+4		+4

41.	42.	43.	44.	45.
4 +5	14 +5	9 −5	19 −5	3 +4

46.	47.	48.	49.	50.
13 +4	2 +5	12 +5	9 −6	19 −6

51.	52.	53.	54.	55.
5 +3	16 +3	4 −3	14 −3	13 −3

Written Exercises

30. Add and check the answers:

1.	2.	3.	4.	5.	6.	7.
3	1	6	9	9	7	2
1	4	0	4	9	8	5
2	7	7	3	1	2	8

8.	9.	10.	11.	12.	13.	14.
6	5	4	9	4	8	3
6	5	3	1	9	8	7
6	4	2	5	4	1	9

Drill Device — A Magic Square

2	7	6
9	5	1
4	3	8

31. 1. Copy the three figures in each row of this square. Three rows run up and down, 3 rows run from left to right, 2 rows from corner to corner. What is the sum of every row? This is called a *magic square*.

2. What figures must be put into the four vacant spaces to give a magic square with 18 as the sum of the three numbers in each row?

5		3
	6	8
9		

If preferred, put this and the following diagrams upon the board for oral work.

3. Supply the six missing numbers, to make a magic square having 10 as the sum of the four figures in each row.

1	2		4
2		1	3
3		4	
	3		1

Counting

32. 1. Frank had 12 hens. He bought 3 more. How many has he now?

2. On Monday he found 4 eggs in one nest and 7 eggs in another. How many did he find in the 2 nests?

3. On Tuesday he found 5 eggs in one nest and 3 in another. How many eggs did he find on Tuesday?

4. How many eggs did he find on Monday and Tuesday?

5. He sold 1 doz. eggs on Wednesday. How many were left?

6. His sister used 3 eggs on Thursday to make a cake. How many had he then?

7. On Friday he found 8 eggs in one nest and 6 in the other. How many did he find in the nests?

8. He sold 6 eggs on Friday. How many had he left?

9. On Saturday he found 8 eggs. How many eggs did he then have?

10. Make other problems like these.

A MEASURE IN COUNTING

12 units or things = 1 dozen (doz.)

Oral Exercise

33. 1. How many splints in Twenty is written 20.

2. How many splints in Twenty-one is written 21.

3. Which figure in 21 splints stands for 1 splint? Which stands for 20 splints?

4. No matter what objects are counted, we may say that in 21 the 1 stands for a *one*, the 2 stands for 2 *tens*.

5. In 22, which figure stands for 2 *ones?* Which stands for 2 *tens?*

6. Read the following numbers:

23 24 25 28 29

7. What does the figure to the right stand for, in 26, 27, 24? In each number, what does the figure to the left stand for?

8. Which is more, 12 or 21?

Counting — Oral

34. 1. Count by twos up to 24, thus: 2, 4, 6, and so on.

2. Count by threes up to 27, thus: 3, 6, 9, and so on.

3. Count by fours up to 29, beginning with 1.

4. Count by fours up to 26, beginning with 2.

5. Count by fours up to 27, beginning with 3.

6. Count by fives up to 26, beginning with 1.

7. Count by 2's from 20 backward.

8. Count by 3's from 21 backward.

Drill Device — The Wheel

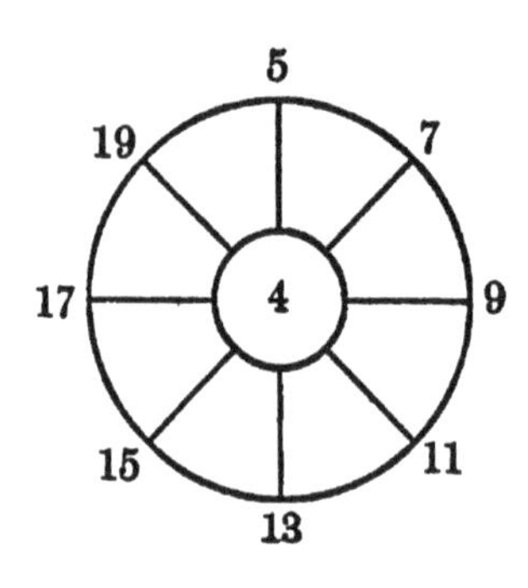

35. 1. In this wheel, add 4 to each of the outside numbers. Learn to do this quickly.

2. Subtract 4 from each of the outside numbers. Use 5 in place of 4. Then add and subtract as before.

3. In the second wheel add 3 to each number on the rim. Subtract 3 from each number.

4. In place of 3 use 4, 5, 6. Add and subtract in each case.

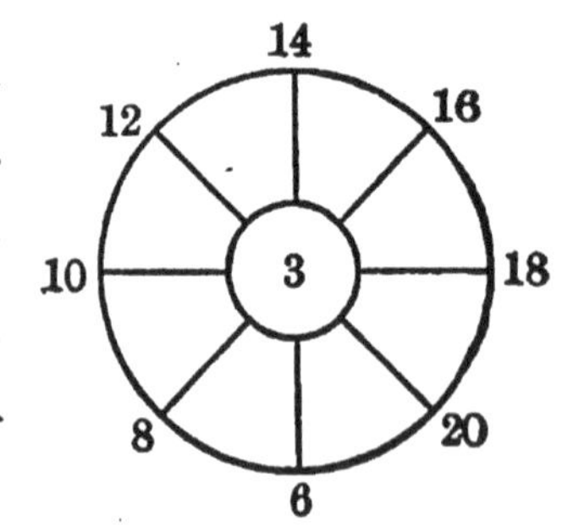

5. Make a drill device of the difficult combination. Drill.

Addition and Subtraction Table

36.

	1	2	3	4	5	6	7	8	9
A	8 7	3 2	8 0	3 3	7 1	7 0	6 4	9 4	7 3
B	9 3	7 7	9 8	7 2	6 2	8 8	8 2	9 7	8 6
C	9 5	6 1	3 1	5 5	3 0	5 4	8 5	5 3	8 1
D	4 2	1 1	7 4	9 6	8 0	7 6	4 3	5 0	6 5
E	5 1	4 1	6 6	6 3	9 1	2 0	9 9	6 0	4 4
F	9 2	4 0	7 5	2 2	5 2	2 1	8 3	9 0	8 4

1. Add at sight the two numbers in each space.
2. Subtract the lower number from the upper in each space.
3. Give answers by lines, then by columns. Do this rapidly.

Exercise in Counting

37. 1. Count by 2's from 30 down to 0.

2. Count by 3's to 27, beginning with 3.
3. Count by 4's to 24, beginning with 4.

4. Count by 5's to 28, beginning with 3.
5. Count by 5's to 27, beginning with 2.
6. Count by 6's to 25, beginning with 1.
7. Count by 6's to 26, beginning with 2.

Written Exercises

38. Write these exercises in numbers, using + for *plus*, and − for *less*:

1. Thirteen *less* three *plus* two = ?
2. Seventeen *less* six *plus* five = ?
3. Twelve *less* seven *plus* four = ?
4. Fourteen *less* nine *plus* five *plus* one = ?
5. Sixteen *plus* one *less* nine *plus* two = ?
6. Eleven *less* seven *plus* two *less* three = ?
7. Find the answers to these exercises.

Written Exercises

39. Write from dictation the following words; write the number after each word:

Eleven	Twenty-one	Twenty-two
Twenty	Sixteen	Nineteen
Ten	Twelve	Twenty-five
Fourteen	Nine	Twenty-four
Thirteen	Fifteen	Twenty-six
Seventeen	Eighteen	Twenty-nine
	Twenty-seven	

Compare the spelling and correct the paper.

MULTIPLICATION AND DIVISION

Multiplication — Oral Exercises

40. 1. How many little squares in *A*? How many little squares of the same size in *B*? How many *A*'s are equal to *B*? How many twos make 4? Write, $2 \times 2 = 4$.

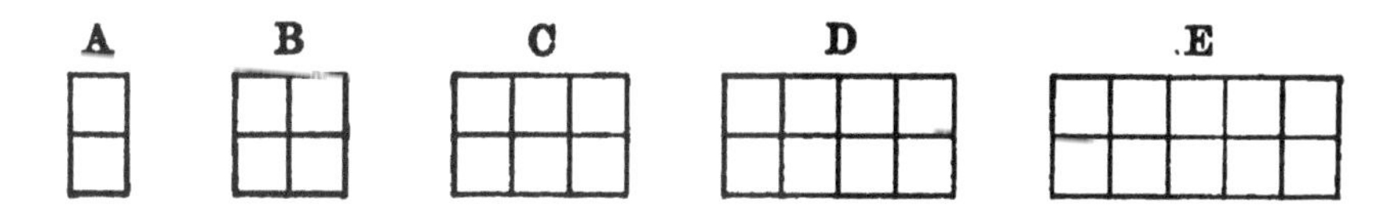

2. How many squares in *C*? How many *A*'s make *C*? How many two's make 6? Write $3 \times 2 = 6$.

3. How many squares in *D*? How many *A*'s make *D*? Write $4 \times 2 = 8$.

4. How many *A*'s make *E*? Hence $5 \times 2 = 10$.

5. Make drawings which show that $6 \times 2 = 12$, $7 \times 2 = 14$, $8 \times 2 = 16$, $9 \times 2 = 18$, $10 \times 2 = 20$.

6. Arrange splints to show that $2 \times 5 = 10$; $5 \times 2 = 10$; $4 \times 3 = 12$; $3 \times 4 = 12$; $2 \times 7 = 14$; $7 \times 2 = 14$; $4 \times 5 = 20$; $5 \times 4 = 20$.

The sign × means *times*. We read 2×2, 2 *times* 2, or 2 *twos*.

41. Memorize the table:

THE MULTIPLICATION TABLE OF 2	
$1 \times 2 = 2$	$6 \times 2 = 12$
$2 \times 2 = 4$	$7 \times 2 = 14$
$3 \times 2 = 6$	$8 \times 2 = 16$
$4 \times 2 = 8$	$9 \times 2 = 18$
$5 \times 2 = 10$	$10 \times 2 = 20$

The multiplication tables should be memorized by means of daily drills. The drill devices given in this book may be used to advantage for this purpose. The order of drill should be varied, namely, $3 \times 2 = 6$. Drill also $2 \times 3 = ?$ $3 \times ? = 6$, etc.

Oral Problems

42. 1. What must you pay for 2 two-cent apples?

2. What is the cost of 3 two-cent stamps?

3. I pay 2¢ apiece for pencils. How many cents will 5 pencils cost?

4. Edith is 2 years old. Julia is 4 times as old. How old is Julia?

5. How much must I pay for 6 two-cent stamps?

6. One piece of ribbon costs 2¢. What will 7 pieces cost?

7. Eight boys are picking nuts. If each boy picks 2 pounds of nuts, how many pounds do they pick?

8. Robert and John each earns $9 a week. How much do they together earn?

9. How much must you pay for 10 two-cent pears?

Oral Problems

43. 1. We see that

▭ and ▭ and ▭ and ▭ = ▭▭▭▭ and ▭▭▭▭

In figures, $4 \times 2 = 2 \times 4$.

2. Make drawings to show that $3 \times 2 = 2 \times 3$.

3. What must you pay for 2 three-cent oranges?

4. How many cents in 1 nickel?

5. How many cents in 2 nickels?

6. How many roses are there in two bunches, if each bunch contains 6 roses?

7. There are 7 days in 1 week. How many days in 2 weeks?

8. At 8¢ a yard, what will 2 yards of braid cost?

9. At $9 a week, how much does a boy earn in 2 weeks?

10. How many cents in two dimes?

11. At $2 a day, what does a man earn in 8 days?

12. Two boys are each 9 years old. What is the sum of their ages?

Drill Device — The Wheel

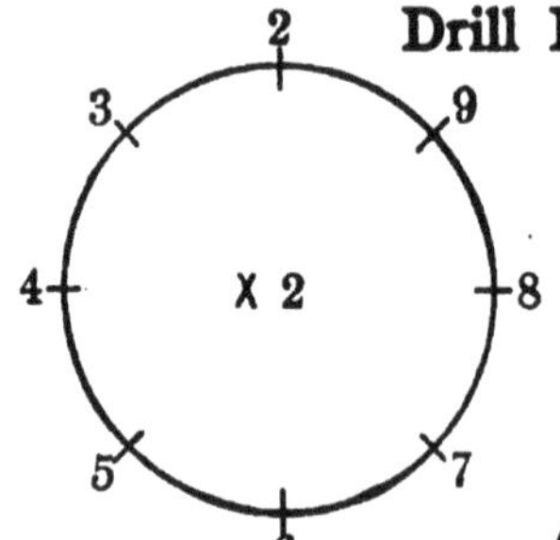

44. Multiply each number on the outside of the wheel by 2.

Change the order of the numbers on the wheel.

Oral Exercises

45. Give the products rapidly:

1.	2.	3.	4.	5.	6.	7.	8.
7	4	3	5	2	6	9	8
× 2	× 2	× 2	× 2	× 2	× 2	× 2	× 2

Liquid Measure — Oral Problems

46. 1. How many pints of milk are there in one quart of milk? In 2 quarts?

1 Pint 1 Quart 1 Gallon

2. How many pint bottles can you fill, if you have 3 quarts of milk?

3. How many pint bottles can be filled from a large can holding 4 quarts of milk?

4. How many times can a pint bottle be filled from a can containing 5 quarts of milk?

5. Mary gets 2 pints of milk each morning and evening. How many quarts does she get a day?

6. A milkman has 6 qt. of milk. How many customers can be supplied with 1 pt. each.

7. How many customers can be supplied with 1 pt. each, from 7 qt. of milk?

8. Make problems on other liquids measured with pint and quart.

MEASURE OF LIQUID

2 pints (pt.) = 1 quart (qt.)
4 quarts = 1 gallon (gal.)

Division — Oral Problems

47. 1. How many 2's are there in 4?

2. How many 2's are there in 6?

3. How many 2's are there in 8?

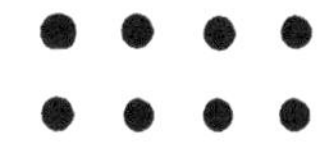

4. How many 2's are there in 10? In 12? In 14? In 18? In 20?

5. A mother divides 14 cherries equally between Mary and James. How many does each get?

6. How many 2-cent pencils can you buy with 18¢?

7. How many pairs of cuffs are there in 16 cuffs?

Division — Oral Exercises

48. Find the missing number:

1. $2 \div 2 = ?$	2. $12 \div 2 = ?$	3. $4 \div 2 = ?$
4. $14 \div 2 = ?$	5. $6 \div 2 = ?$	6. $16 \div 2 = ?$
7. $8 \div 2 = ?$	8. $18 \div 2 = ?$	9. $10 \div 2 = ?$
10. $20 \div 2 = ?$		

$\begin{array}{r} 8 \\ 2\overline{)16} \end{array}$ means that 2 is contained in 16, 8 times. It may also be written $16 \div 2 = 8$. The $\overline{)\quad}$ or $\div$ indicates division. The answer is called the **quotient**.

49. Study the table:

THE DIVISION TABLE OF 2

$2 \div 2 = 1$	$\begin{array}{r} 1 \\ 2\overline{)2} \end{array}$	$12 \div 2 = 6$	$\begin{array}{r} 6 \\ 2\overline{)12} \end{array}$
$4 \div 2 = 2$	$\begin{array}{r} 2 \\ 2\overline{)4} \end{array}$	$14 \div 2 = 7$	$\begin{array}{r} 7 \\ 2\overline{)14} \end{array}$
$6 \div 2 = 3$	$\begin{array}{r} 3 \\ 2\overline{)6} \end{array}$	$16 \div 2 = 8$	$\begin{array}{r} 8 \\ 2\overline{)16} \end{array}$
$8 \div 2 = 4$	$\begin{array}{r} 4 \\ 2\overline{)8} \end{array}$	$18 \div 2 = 9$	$\begin{array}{r} 9 \\ 2\overline{)18} \end{array}$
$10 \div 2 = 5$	$\begin{array}{r} 5 \\ 2\overline{)10} \end{array}$	$20 \div 2 = 10$	$\begin{array}{r} 10 \\ 2\overline{)20} \end{array}$

The division table should be taught in connection with the multiplication table. Daily drill should be given to fix the combinations in mind.

Drill Device — The Wheel

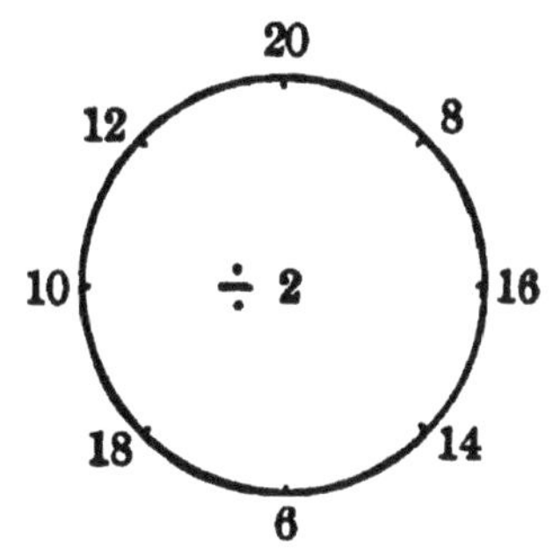

50. Divide each number on the outside of the wheel by 2. Change the order of the numbers on the wheel.

Oral Problems

51. 1. How many two-cent postage stamps can be bought with 10¢? With 14¢? With 18¢? With 12¢? With 16¢? With 20¢?

2. John buys 3 two-cent stamps and 4 one-cent stamps. How much does he pay for the stamps?

3. Another time he buys 9 two-cent stamps. How much change does he receive from 20¢?

4. How many quart bottles can be filled from a can holding 18 pints of milk?

5. How many shoes are there in 7 pairs of shoes?

6. A farmer has 16 horses. How many teams of two horses each has he?

7. How many gloves are there in 9 pairs of gloves?

8. Mrs. Ross uses one quart of milk every morning, and one quart of milk every evening. How many quarts of milk does she use in a week?

9. Edward is 7 years old and Charles is twice as old. How old is Charles?

10. James is twice as old as Ruth. James is 12 years old. How old is Ruth?

11. Frank walks 4 miles an hour. How far can he walk in 2 hours?

12. How long will it take to go 10 miles at the rate of 2 miles an hour?

Review — Oral

52. Find the answer:

1. 9×2	2. 2×6	3. 7×2	4. 5×2	5. 8×2
6. $2\overline{)20}$	7. $2\overline{)18}$	8. $2\overline{)14}$	9. $2\overline{)8}$	10. $2\overline{)10}$
11. $\begin{array}{r}15\\+5\\\hline\end{array}$	12. $\begin{array}{r}17\\-9\\\hline\end{array}$	13. $\begin{array}{r}13\\-7\\\hline\end{array}$	14. $\begin{array}{r}17\\+3\\\hline\end{array}$	15. $\begin{array}{r}13\\+6\\\hline\end{array}$
16. $\begin{array}{r}20\\-7\\\hline\end{array}$	17. $\begin{array}{r}18\\-9\\\hline\end{array}$	18. $\begin{array}{r}13\\+4\\\hline\end{array}$	19. $\begin{array}{r}14\\+3\\\hline\end{array}$	20. $\begin{array}{r}15\\+4\\\hline\end{array}$
21. $\begin{array}{r}15\\-8\end{array}$	22. $\begin{array}{r}19\\-6\end{array}$	23. $\begin{array}{r}18\\-7\end{array}$	24. $\begin{array}{r}12\\-6\end{array}$	25. $\begin{array}{r}16\\-8\end{array}$

Multiplication by Two

53. 1. Multiply 14 by 2.

PROCESS

$$\begin{array}{r} 14 \\ \times 2 \\ \hline 28 \end{array}$$

EXPLANATION.—Write the 2 beneath the 4 in 14.

$2 \times 4 = 8$. Write the 8 under the 4.

$2 \times 1 = 2$. Write the 2. The answer, 28, is the *product*.

2. If a yard of lace costs 12¢ and a yard of ribbon 5¢, how much must I pay for 2 yards of lace and 1 yard of ribbon?

PROCESS

$$\begin{array}{r} 12¢ \\ \times 2 \\ \hline 24¢ \end{array}$$

24¢ + 5¢ = 29¢

EXPLANATION.—Find the product of the two numbers, 12 and 2. Add to find the answer.

3. A storekeeper sold 2 overcoats, on account, at $13 each. On one of these coats a man paid $5. How much is still due on both coats?

PROCESS

$$\begin{array}{r} \$13 \\ \times 2 \\ \hline \$26 \end{array}$$

$26 − $5 = $21

EXPLANATION.—Find the product of the two numbers, 13 and 2. Subtract to find the difference.

Written Problems

54. 1. There are 2 pints in 1 quart. How many are there in 13 quarts?

2. How many apples are 2 dozen apples?

3. A gallon of oil costs 11¢. What will 2 gallons cost?

4. If 1 pencil costs 2¢, what will a dozen pencils cost at the same rate?

5. A man works 10 hours a day. How many hours does he work in 2 days?

6. A boy buys 5 two-cent oranges and a one-cent banana. How much must he pay?

7. John earns 10¢ a day for 2 days. He spends 4¢ for candy. How much has he left?

8. A party took 2 dozen apples to a picnic. They ate 20 apples and gave the rest away. How many apples were given away?

9. Ralph earns $14 a month and $— in 2 months?

10. Robert earns $2 a week carrying papers. How much does he earn in 13 weeks?

11. A camping party travels 11 miles a day. How far does it go in 2 days?

Division — Written Exercises

55. 1. Divide 26 by 2.

PROCESS

$$\begin{array}{r} 13¢ \\ 2\overline{)26¢} \end{array}$$

EXPLANATION. — Make the division sign as indicated.

$2 \div 2 = 1$. Write the 1 over the 2 for the first figure in the answer.

$6 \div 2 = 3$. Write the 3 over the 6 for the second figure in the answer.

The answer, 13¢, is the *quotient*.

2. $2\overline{)28}$ 3. $2\overline{)22}$ 4. $2\overline{)24}$ 5. $2\overline{)20}$

Written Problems

56. 1. How many quarts are there in 26 pints?

2. How many 2-cent stamps can be purchased with 26¢?

3. There are 22 trees in 2 rows of equal length. How many trees in each row?

4. There are 28 buttons on a card. They are arranged in 2 rows of equal number. How many buttons in each row?

5. A class of 24 pupils can be seated on 2 recitation benches each holding the same number of pupils. How many does each bench seat?

Division by Two — Halves

57. 1. Into how many parts is this apple divided?

2. How do these parts compare in size?

3. What is each part called?

4. How many halves are there in the whole apple?

5. Into how many parts is this sphere divided?

6. What is each part called?

7. How many halves are there in the whole sphere?

8. Into how many equal parts must an apple or a sphere be divided to get one half of it?

To find **one half** of a number, divide the number by **2**.

One half is written $\frac{1}{2}$

Oral Problems

58. 1. A boy had 14¢ and spent half of it for a pad and a pencil. How much money has he left?

2. A blackboard is 12 ft. long. What is one half its length?

3. There are 10 pupils in a class. How many pupils are there in half the class?

4. Frank earned 16 stars for perfect work. Andrew earned half as many. How many did Andrew earn?

5. A boy can do a piece of work in 10 da. His father can do it in half the time. How long will it take his father to do the work?

Oral Exercises

59. Give quickly $\frac{1}{2}$ of each of the following.

1. 12. 2. 18. 3. 10. 4. 8. 5. 14. 6. 16.

Give the answers:

7. $2\overline{)4}$. 8. $2\overline{)6}$. 9. $2\overline{)10}$. 10. $2\overline{)12}$.

11. $\frac{1}{2}$ of 4. 12. $\frac{1}{2}$ of 6. 13. $\frac{1}{2}$ of 10. 14. $\frac{1}{2}$ of 12.

Drill Device — The Wheel

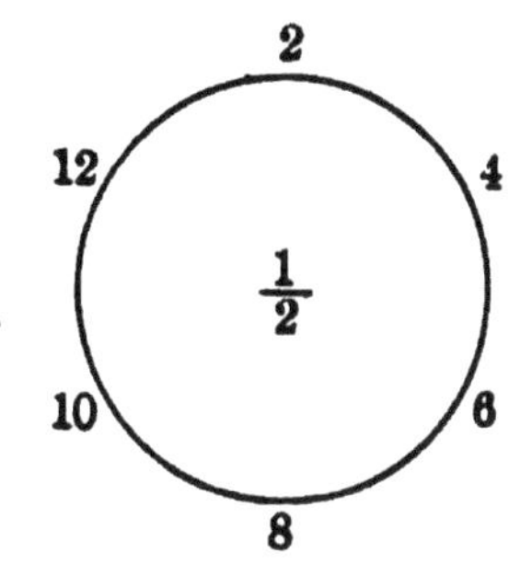

60. Give $\frac{1}{2}$ of each number on the rim of the wheel.

Change the order of the numbers on the rim of the wheel.

Comparison and Measurement — Exercises

61. 1. Line *A* is 1 in. long.

2. Point to line *B*. About how long is it?

C

D

B

F

A

E

3. About how long is line *C*?

4. Write the length of line *B* in figures.

5. Write the length of line *C* in figures.

6. How long is line *D*? Write the length in figures.

7. How long is line *F*? Write the answer in figures.

8. Compare the length of line *F* with line *A*. Which is longer?

9. How many times can line *F* be marked off on line *A*? Write the answer.

10. Compare the length of line *B* with line *D*.

11. Compare the length of line *F* with line *B*.

12. Compare the length of line *C* with line *E*.

13. Measure the lines with a foot rule.

Measuring

62. 1. A boy with a yardstick measures the length of a fence. He makes a chalk mark at the end of each yard length. How many yards long is the fence?

2. A girl measures the height of the fence with a foot rule. How high is the fence?

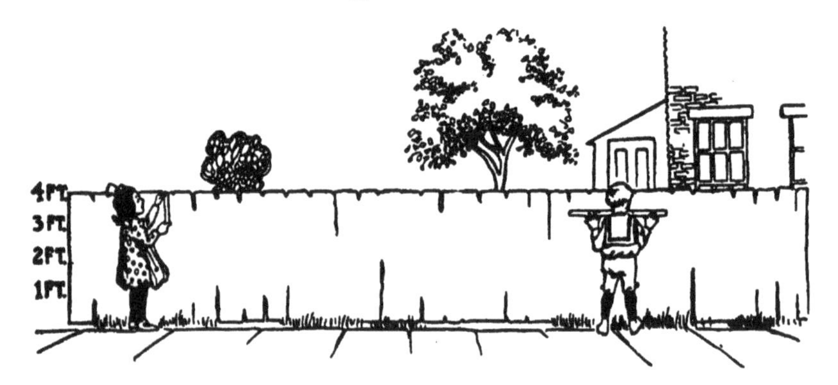

3. Is the height of the fence more than 1 yd.? How much more?

Review — Oral Problems

63. 1. An errand boy earned 5¢ on Monday, 3¢ on Tuesday, and 7¢ on Wednesday. How many cents did he earn in the three days?

2. There were 16 quails in a flock. 8 of them flew away. How many quails are left?

3. Jane and Kate had 18 cherries. Kate ate 9. How many cherries did Jane eat?

4. A milkman has 6 quarts of milk. He sells a pint of milk to each customer. How many customers can he supply?

5. How many pints in 5 quarts? In 4 quarts? In 7 quarts?

6. At 2¢ apiece, how much will 8 oranges cost? 3 oranges? 9 oranges?

7. Charles picked 16 cherries. He gave 7 to George and 5 to his sister. How many cherries had he left?

8. Dan had 18 marbles. He lost 10 and gave away 3. How many had he left?

9. If one lemon costs 2¢, how many lemons can you get for 20¢?

10. A girl had 11¢. She bought 2 yd. of ribbon at 4¢ a yard. How much money has she left?

11. A boy has 15¢ and buys 4 oranges at 2¢ each. How much money has he left?

12. A woman had 25 ¢. She paid 6 ¢ for a box of crackers and 5 ¢ for a bottle of milk. How much change did she receive?

13. What is the cost of 2 rubber balls at 5 ¢ each, a top at 6 ¢, and a whistle at 10 ¢?

14. How many blocks at 2 ¢ each can be bought for 18 ¢?

15. Mary has 25 ¢. She buys 2 yd. of ribbon at 8 ¢ a yard. How much money has she left?

16. How many tin soldiers at 2 ¢ each can be bought for 18 ¢?

17. How many pints are there in 8 quarts and 1 pint?

18. A cook uses a pint of molasses in one baking. How many bakings can she make with 9 qt. and 1 pt.?

19. A quart of ice cream is served to 6 people. How many people can be served with 4 pints of ice cream?

Problems and Exercises — Oral

64. 1. What is the largest number that can be written with one figure?

2. How many figures are needed to write any one number between *ten* and *twenty?* What does each digit stand for in 16, 19, 25?

3. Here are two groups of splints.

How many splints are tied in each bundle?

4. Do you see in the left-hand group 10 splints and 2 splints? How much is $10 + 2$?

5. In the second group, how many bundles of 10 splints each? How many splints are left over? How many splints in the second group?

6. How many bundles and how many extra splints would stand for the number 17? For the number 25? For the number 36?

7. In 36, how many *ones* does the figure to the right stand for? How many *tens* does the figure to the left stand for?

8. Tell what each figure stands for in the following numbers: 31, 13, 23, 32, 44.

9. Count from 30 to 50.
10. Count from 50 to 80.
11. Count from 80 to 100.

Money

65. 1. How many cents in a ten-cent piece (or dime)?

2. How many cents are equal to 1 ten-cent piece and 7 cents? Write the number of cents.

3. How many cents are equal to 2 ten-cent pieces and 5 cents?

4. How many cents are equal to 3 ten-cent pieces and 1 cent?

The School Store

Appoint a storekeeper. Let the other children be the customers. Use toy money. Direct the purchases and correct the making of change. It is not necessary that the articles purchased be real. Let the children use their imagination and suggest things to be bought.

Drill Exercise

67. What does each figure stand for in the following numbers:

56 = 5 tens + 6 ones

87	78	64	46	95	59	66
34	28	90	19	81	72	63

Written Exercise

68. Write the following in numbers:

Forty-six	Fifty	One hundred
Nineteen	Twenty-eight	Ninety-nine
Thirty-seven	Sixty-one	Sixty-five
Seventy	Seventy-two	Thirty-one
Forty-nine	Ninety-four	Eighty
Fifteen	Twenty-four	Eighty-eight
Seventy-seven	Forty-five	Fifty-five

ADDITION AND SUBTRACTION

Counting by Decades

69. 1. Count by 2's from 20 to 40.

2. Count by 2's from 30 to 10.
3. Begin with 50 and count by 2's to 70.
4. Start with 30 and count by 3's to 51.
5. Start with 61 and count by 3's to 79.
6. Begin with 81 and count by twos to 99.
7. Begin with 81 and count by threes to 96.
8. Count from 45 to 30.
9. Count by threes from 21 to 36.
10. Count by threes from 27 to 0.
11. Begin with 13 and count by fives to 50.
12. Begin with 16 and count by fives to 51
13. Count by fives from 40 to 0.
14. Count from 43 to 29.
15. Count by twos from 48 to 28.
16. Take away 4 from 17, 47, 37, 57, 67, 97.
17. Take away 3 from 15, 75, 65, 55, 45, 25.
18. Take away 6 from 19, 79, 89, 99, 49.
19. Add 7 to 14, 34, 54, 94, 74, 84.
20. Add 5 to 28, 58, 38, 88, 98.
21. Add 9 to 12, 32, 72, 62, 42, 22.

Review by using drill devices for review in counting.

Written Exercises

70.

1. 19 +5	2. 29 +5	3. 99 +5	4. 79 +5	5. 59 +5	6. 49 +5
7. 23 +6	8. 53 +6	9. 66 +6	10. 36 +6	11. 35 +6	12. 75 +6
13. 31 −3	14. 51 −3	15. 41 −3	16. 42 −3	17. 62 −3	18. 82 −3
19. 13 −8	20. 73 −8	21. 53 −8	22. 44 −8	23. 64 −8	24. 94 −8

Subtraction by the Austrian Method

Where the figure in the minuend is less than the corresponding figure in the subtrahend, the process is as follows:

$$\begin{array}{r} 42 \\ -3 \\ \hline 39 \end{array}$$

PROCESS

$$\begin{array}{r} 4^{1}2 \\ 3 \\ \hline 39 \ \textit{Ans.} \end{array}$$

EXPLANATION. The 2 is less than the 3.

From 4 *tens* take away 1 *ten* and add it to 2 *ones*. This gives 12 *ones*.

Then 3 *ones* and 9 *ones* are 12 *ones*. Write the 9.

In the tens column 3 *tens* are left over. Write the 3. The answer is 39.

Written Exercises

71. Write in figures:

1. The numbers from 30 to 40.
2. The numbers from 50 to 60.
3. The numbers sixty-five, fifty-six, eighty-one.
4. The numbers twenty-six, eighty-five, fifty.

Written Exercises

72. Write in words:

1.	54,	64,	74,	84,	94,	4,	60.
2.	53,	23,	63,	33,	73,	43,	83.
3.	90,	20,	80,	30,	70,	40,	50.
4.	55,	66,	77,	88,	99,	44,	33.
5.	87,	65,	81,	72,	93,	45,	67.

Oral Exercises

73. Add by decades to 100:

1.	11	21	31	41	51	61
	1	1	1	1	1	1
2.	12	22	32	42	52	62
	2	2	2	2	2	2
3.	13	23	33	43	53	63
	3	3	3	3	3	3
4.	17	27	37	47	57	67
	7	7	7	7	7	7

5.	18	28	38	48	58	68
	8	8	8	8	8	8
6.	29	39	49	59	69	79
	6	6	6	6	6	6

Addition

74. 1. Add 25 and 38.

PROCESS

tens	ones
2	5
3	8
6	3 *Ans.*

EXPLANATION. Adding the ones: $8 + 5 = ?$ ones. 13 ones = 1 ten and 3 ones. Write the 3 in the column of ones. Then carry the 1 ten to the column of tens. Adding *tens:* 1 ten + 3 tens + 2 tens = ? tens. Write the 6 tens in the column of tens.

The answer, 63, is the *sum.*

To make sure that there is no mistake in the answer, add the numbers again, but add *downward.* In this way *check* the answer.

Written Exercises

75. Add the following:

1. 24	2. 34	3. 65	4. 53	5. 28	6. 39
27	57	16	29	34	44
7. 77	8. 42	9. 49	10. 38	11. 57	12. 46
14	19	29	48	27	26

13. 33¢	14. 12 lb.	15. 21 ft.	16. 53 in.	17. $46
48¢	48 lb.	19 ft.	27 in.	$24

Written Problems

76. 1. A boy earned 23 cents one day, and 28 cents the next day. How much did he earn in the two days?

2. One day 27 apples fell from one tree and 35 from another. How many apples fell from the two trees?

3. There are 35 pupils in one class and 27 in another. How many pupils in the two classes?

4. Mr. Stuart had $ 64 and earned $ 27 more. How much has he now?

5. Louise has 58 roses and 45 lilies in her garden. How many flowers has she?

6. A sidewalk in front of a man's house was 47 ft. long. Twenty-six feet more were laid. How many feet of sidewalk are there now?

7. In a school there are 37 boys and 45 girls. How many pupils are there in the school?

8. William had 35¢ in the bank and put in 18¢ more. How much money has he in the bank?

The Fruit Stand — Practical Applications

77. At this fruit stand prices are as follows:

Apples, 3 for 5¢ Oranges, 2 for 5¢.
Bananas, 1 for 2¢ Cherries, 10¢ a pound.

1. What is the cost of 6 apples? 5 bananas? 2 lb. of cherries? 4 oranges?

2. A boy buys 3 lb. of cherries and gives the clerk 50¢. How much change should he receive?

3. Another boy gets 6 apples and 2 bananas. How much change does he receive from a quarter?

4. If you buy 7 bananas, 1 lb. cherries, and 3 oranges, what must you pay?

5. Appoint a keeper for the stand. Direct the pupils in their purchases. Use toy money. Let change be made in the usual manner. Let a pupil write the transactions on the blackboard.

READING AND WRITING NUMBERS TO 1000

Introductory Exercise

78. 1. What is the largest number you can write with two figures?

2. What is the smallest number you can write with three figures?

One hundred is written 100 *One thousand* is written 1000

3. How many ones make 1 ten?
4. How many tens make 1 hundred?
5. How many hundreds make 1 thousand?
6. How many ten-cent pieces make 100 cents?
7. How many ten-cent pieces make 200 cents?
8. Write one hundred forty in figures.
9. Write one hundred ten in figures.
10. Write one hundred one in figures.

368 = 3 hundreds + 6 tens + 8 ones
947 = 9 hundreds + 4 tens + 7 ones
802 = 8 hundreds + 0 tens + 2 ones

Read thus:

325, three hundred twenty-five
860, eight hundred sixty
905, nine hundred five

Oral Exercises

79. Tell whether the figures in the following numbers stand for thousands, hundreds, tens, or ones:

1.	29	2.	67	3.	100	4.	300	5.	591
6.	40	7.	168	8.	205	9.	289	10.	1000

Reading Numbers — Oral

80. Read the following:

1.	2.	3.	4.	5.	6.
211	101	218	121	102	201
13	31	103	301	214	241
401	104	28	82	802	820
810	208	280	140	410	769
119	91	910	190	901	109
876	67	706	607	670	760

Written Exercise

81. Write in figures:

1. Nineteen. Sixty-three. Nine hundred nine.
2. Eighty. Twenty-nine. Five hundred fifty.
3. Seventy. Seventy-two. Seven hundred seventeen.
4. Twelve. Five hundred. Five hundred twelve.
5. Fifty. One hundred. Four hundred four.
6. Ninety. Seven hundred. Two hundred nine.

7. Seventeen. Two hundred. Three hundred six.
8. Sixty. Six hundred. Five hundred ninety.
9. Eighty-one. Nine hundred. Three hundred ten.
10. Seventy-five. Three hundred. Three hundred seventy-five.
11. Twenty-one. Four hundred. Eight hundred sixty-nine.
12. Ninety-nine. One thousand. Seven hundred twenty-three.
13. Seventy-seven. Four hundred forty. One hundred ninety-nine.
14. Forty-nine. Nine hundred forty. Two hundred forty-eight.
15. Thirty-six. Five hundred nineteen. Four hundred eighty-nine.

82. Write in words:

1.	2.	3.	4.	5.	6.
10	11	12	13	14	15
16	17	18	19	20	25
29	30	40	90	110	120
200	250	219	225	790	1000

83. Find the sum and write in figures and words:

1. $10+5$ 2. $20+3$ 3. $40+7$ 4. $50+1$
5. $70+6$ 6. $80+7$ 7. $90+2$ 8. $100+10+2$

9. $700+40+5$ 10. $600+70+3$

11. $300+20+8$ 12. $200+80+5$

One dollar and twenty-five cents is written \$1.25

Two dollars and forty-five cents is written \$2.45

The point (.) is used to separate the dollars from the cents.

Money — Written Exercises

84. Write in figures:

1. Three dollars and fifteen cents.
2. Ten dollars and fifty cents.
3. Five dollars and ninety-five cents.
4. One dollar and seventy-five cents.
5. Two dollars and ten cents.
6. Three hundred dollars.

85. Find the sum and write in figures and words:

1. \$400 + \$70 + \$5 2. \$200 + \$10 + \$500

3. \$300 + \$10 + \$2 4. \$100 + \$30 + \$6

5. \$500 + \$6 6. \$600 + \$70

A Number Game

86. Toss bean bags into the circles.

A bean bag stopping in circle A counts 5.

A bean bag stopping outside of A but inside of B counts 3.

A bean bag stopping outside of B but inside of C counts 1.

Each player is allowed six chances.

SCORE

	INSIDE A	INSIDE B ONLY	INSIDE C ONLY
Jane	twice	once	3 times
Mary	once	twice	3 times
Ada	twice	twice	twice

Jane's score is

$$2 \times 5 + 1 \times 3 + 3 \times 1 = 10 + 3 + 3 = 16$$

1. Find Mary's score.
2. Find Ada's score.

3. Find the total of the three scores.

4. Continue and vary the form of the exercises.

Exercise in Counting

87. Count:

1. By 5's from 60 to 80.
2. By 3's from 42 to 99.
3. By 4's from 70 to 90.
4. By 5's from 45 to 95.
5. By 3's from 50 to 71.
6. By 4's from 44 to 100.
7. By 6's from 70 to 100.
8. By 6's from 36 to 60.

ADDITION

88. Add rapidly:

1.	2.	3.	4.	5.	6.	7.
10	11	9	10	11	9	3
5	3	3	12	15	10	13
2	5	4	13	21	30	12

8.	9.	10.	11.	12.
4	11	20	30	60
5	10	3	40	10
10	10	10	5	1
10	4	5	1	7

Oral Problems

89. 1. Mary has 10¢, John 25¢, Ralph 10¢, Lucy 20¢. How much have they together?

2. It costs 50¢ to hire a boat. James has 20¢, George a dime, Ned 5¢, Frank 13¢. How much must they borrow?

3. Will has 15 books, Henry 10 books, James 11 books, Ralph 4 books. They agree to put all their books into a joint library. How many books will there be?

4. There are 11 boys and 13 girls in a class. How many pupils are there in the class?

5. Mrs. Smith had 75¢ in her purse. She gave her daughter 25¢ and spent 10¢ for carfare. How much has she left?

6. The distance between Rochester and Niagara Falls is 78 mi. A man started from Rochester and walked 18 mi. the first day and 10 mi. the second day. How far was he from Niagara Falls at the end of the second day?

Oral Problems

90. 1. James spent 40 cents for a bat, 30 cents for a ball, and 10 cents for an old glove. How much did he spend in all?

2. A man traveled in an automobile 45 miles one day, 9 miles the next day, and 8 miles the third day. How many miles did he travel in the three days?

3. A grocer had 20 watermelons. He sold 3, then 2, then 4. How many were left?

4. A boy had 83¢ in his bank. After spending 8¢ he put 5¢ more in the bank. How much has he in the bank?

5. A boy had 60¢ to spend for seed. He pays 10¢ for a package of radish seed, 10¢ for lettuce seed, and 25¢ for grass seed. How much has he left?

The Fence — Drill Device

91. Give the answers from addition.

Give the answers from subtraction.

Change the numbers and the order of the numbers.

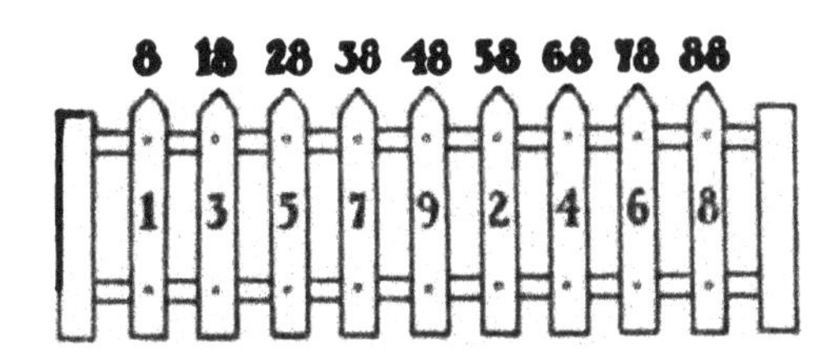

Addition — Carrying

92. 1. Add 15, 36, 23.

PROCESS

tens ones
15
36
23
74 *Ans.*

EXPLANATION. — Adding:

3 ones + 6 ones + 5 ones = ? ones.

14 ones = 1 ten and 4 ones.

Write the 4 ones in the column of ones.

Then carry the 1 ten to the column of tens.

Add the tens.

1 ten + 2 tens + 3 tens + 1 ten = ? tens.

Write the 7 tens in the column of tens.

74 is the sum. Check the answer.

2. 43 + 23 + 18 = ?

PROCESS

43
23
18
84 *Ans.*

EXPLANATION. — Write the numbers in a column, ones under ones, and tens under tens. Then add. Check the answer.

Written Exercises

93. Find the sum. Check the answer.

1.	2.	3.	4.	5.	6.
19	23	56	12	25	70
21	34	27	22	18	13
15	16	11	37	19	9

7.	8.	9.	10.	11.	12.
38	65	42	71	53	32
41	14	37	24	45	25
11	11	11	15	12	13

13.	14.	15.	16.	17.
17¢	27¢	36¢	58¢	46¢
26¢	19¢	18¢	17¢	13¢

18.	19.	20.	21.
$3.10	$11.60	$ 7.30	$13.15
4.75	2.25	13.50	14.50

Written Exercises

94. Write the answers:

1. 30 + 40 + 19
2. 70 + 16 + 22
3. 34 + 43 + 19
4. 10 + 12 + 73
5. 13 + 11 + 27 + 13
6. 14 + 74 + 10
7. 10 + 10 + 9 + 11
8. 17 + 71 + 49
9. 90 + 80 + 77 + 66
10. 55 + 44 + 33
11. 20 + 30 + 15 + 11
12. 55 + 77 + 88
13. 99 + 88 + 70 + 60
14. 55 + 66 + 77

SUBTRACTION

95. Count:

1. By 10's from 100 back to 0.
2. By 5's from 100 back to 50.
3. By 2's from 100 back to 80.
4. By 3's from 18 back to 0.
5. By 4's from 64 back to 24.
6. Change the larger numbers.

Drill Device — The Circle

96. Subtract the number in the center of the circle from the number on the outside of the circle.

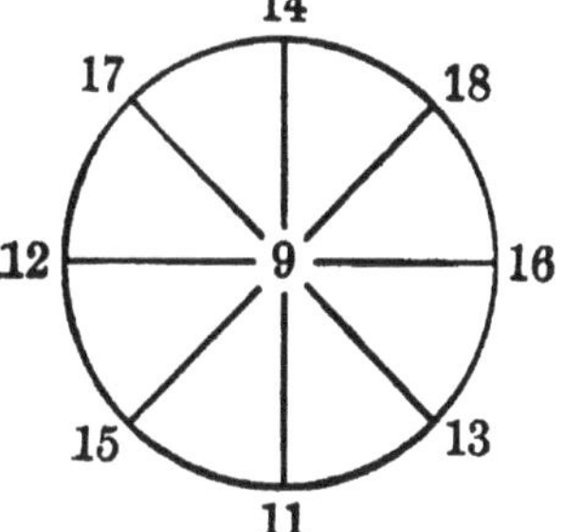

Change the order of the numbers on the outside.

Change the numbers on the outside.

Change the number in the center so as to use 7, 6, etc.

Written Exercises

97. 1. Subtract 25 from 59.

PROCESS

tens ones
 59
−25
 34 *Ans.*

EXPLANATION. — Write 25 below 59, ones under ones and tens under tens. 9 ones − 5 ones = ? 5 tens − 2 tens = ?

2. 44	3. 64	4. 57	5. 48	6. 39	7. 28
−22	−23	−22	−15	−25	−17
8. 90	9. 99	10. 86	11. 78	12. 62	13. 29
−70	−71	−52	−43	−32	−12
14. 74	15. 87	16. 92	17. 39	18. 46	19. 55
−33	−45	−62	−20	−31	−32
20. 23	21. 44	22. 55	23. 70	24. 90	25. 85
−11	−22	−22	−50	−70	−25

Written Problems

98. 1. A farmer has 95 cows. He sells 42 of them. How many has he left?

2. Mary has read 53 pages in a book of 78 pages. How many pages has she still to read?

3. A train runs from Milwaukee to Chicago, a distance of 85 miles. How far is the train from Chicago when it is 42 miles from Milwaukee?

4. A commercial school has 76 typewriters. Fourteen are not in use. How many are in use?

5. A man bought 8 tons of coal for $48. He paid $25 on account. How much does he owe?

6. A boy sold 84 papers last week and 62 the week before. How many more did he sell last week?

7. A woman spends 55¢ for a pound of tea and 35¢ for a pound of coffee. How much more does she pay for the tea?

8. A fruit dealer had 76 oranges and sold 32. How many are left?

Subtraction — Borrowing

99. 1. Subtract 6 from 85.

PROCESS

tens	ones
8	5
−	6
7	9 *Ans.*

EXPLANATION. — It is impossible to take 6 ones from 5 ones.

From the 8 tens take 1 ten, leaving 7 tens.

Add 1 ten to the 5 ones, making 15 ones.

Then, 6 ones and 9 ones make 15 ones. Write the 9 under the ones.

Bring down the 7 tens.

The answer is 79.

2. Subtract 8 from 73.

PROCESS

tens	ones
7	3
−	8
6	5 *Ans.*

EXPLANATION. — Take 1 ten from the 7 tens, leaving 6 tens.

Add the 1 ten to the 3 ones, making 13 ones.

8 ones and 5 ones make 13 ones. Write the 5 under the ones. Bring down the 6 tens. The answer is 65.

3. Subtract 9 from 64.

PROCESS

tens	ones
6	4
−	9
5	5 *Ans.*

EXPLANATION. — Take 1 ten from the 6 tens, leaving 5 tens.

Add the 1 ten to the 4 ones, making 14 ones. 9 ones and 5 ones make 14 ones.

Write the 5 under the ones. Bring down the 5 tens. The answer is 55.

Written Exercises

100. Subtract:

1. 42 −5	2. 52 −5	3. 73 −5	4. 64 −5	5. 82 −4	6. 92 −4
7. 75 −6	8. 85 −6	9. 71 −6	10. 51 −6	11. 63 −6	12. 83 −6
13. 95 −7	14. 55 −7	15. 64 −7	16. 84 −7	17. 81 −7	18. 71 −7
19. 43 −8	20. 53 −8	21. 74 −8	22. 84 −8	23. 95 −8	24. 65 −8
25. 97 −9	26. 58 −9	27. 76 −9	28. 65 −9	29. 84 −9	30. 73 −9
31. 62 −9	32. 71 −9	33. 80 −9	34. 91 −7	35. 82 −8	36. 93 −7

SUBTRACTION — TENS AND ONES

For Study and Discussion

101. 1. Subtract 28 from 64.

PROCESS

tens ones	
64	
− 28	
36	*Ans.*

EXPLANATION. — Take 1 ten from the 6 tens, leaving 5 tens.

Add the 1 ten to the 4 ones, making 14 ones.

8 ones and 6 ones make 14 ones. Write the 6 under the ones.

Then, 2 tens and 3 tens make 5 tens. Write the 3 under tens.

The answer is 36.

Written Exercises

102. Subtract:

1. 31 − 12	2. 31 − 13	3. 31 − 14	4. 31 − 15	5. 31 − 16
6. 32 − 13	7. 32 − 14	8. 32 − 15	9. 32 − 16	10. 32 − 17
11. 35 − 16	12. 35 − 17	13. 35 − 18	14. 35 − 19	15. 45 − 19
16. 37 − 18	17. 42 − 17	18. 65 − 26	19. 38 − 29	20. 23 − 15

Written Exercises

103. Find the missing number:

1.	2.	3.	4.	5.	6.
27	35	23	41	42	83
−16	−?	−15	−18	−?	−14
?	24	?	?	27	?

7.	8.	9.	10.	11.	12.
75	63	83	?	?	?
−16	−?	−?	−28	−35	−36
?	43	58	47	28	47

Written Problems

104. 1. There are 15 cows in one field and 29 cows in another field. How many more cows are there in the second field than in the first?

2. A farmer has 35 acres of wheat and 26 acres of corn. How many more acres of wheat than corn has he?

3. A man started on a bicycle to reach a railroad station, 36 miles distant. When he had ridden 17 miles, he stopped at a hotel for dinner. How many miles was he from the station?

4. A man earns $65 a month and saves $27 of this sum. What are his monthly expenses?

5. Harry is 13 years old and his father is 41. How much older is the father?

6. Charles has 38¢ and George has 57¢. How many more cents has George than Charles?

7. A boy, carrying a basket with 43 eggs, fell and broke 27 of them. How many eggs were not broken?

8. There are 27 girls in one class and 35 in another. How many girls in both classes? How many more in the second class?

9. John is handed 50¢ and asked to purchase 32¢ worth of stamps. How much change should John bring back?

10. 87 boys and 69 girls were enrolled in a school. Find the difference in the number of boys and girls.

Subtraction — Ones, Tens, Hundreds, Thousands

105. Subtract:

PROCESS AND EXPLANATION

hundreds tens es

1.
$$\begin{array}{rl} 250 & = 2 \text{ hundreds} + 5 \text{ tens} + 0 \text{ ones} \\ -124 & = 1 \text{ hundred} + 2 \text{ tens} + 4 \text{ ones} \\ \hline 126 & \end{array}$$

or

$$\begin{array}{rl} 250 & = 2 \text{ hundreds} + 4 \text{ tens} + 10 \text{ ones} \\ -124 & = 1 \text{ hundred} + 2 \text{ tens} + 4 \text{ ones} \\ \hline 126 & = 1 \text{ hundred} + 2 \text{ tens} + 6 \text{ ones} \end{array}$$

2.
$$\begin{array}{rl} 348 & = 3 \text{ hundreds} + 4 \text{ tens} + 8 \text{ units} \\ -159 & = 1 \text{ hundred} + 5 \text{ tens} + 9 \text{ units} \\ \hline 189 & \end{array}$$

or

$$\begin{array}{rl} 348 & = 2 \text{ h.} + 14 \text{ t.} + 8 \text{ o.} = 2 \text{ h.} + 13 \text{ t.} + 18 \text{ o.} \\ -159 & = 1 \text{ h.} + 5 \text{ t.} + 9 \text{ o.} = 1 \text{ h.} + 5 \text{ t.} + 9 \text{ o.} \\ \hline 189 & \qquad\qquad\qquad\qquad\quad = 1 \text{ h.} + 8 \text{ t.} + 9 \text{ o.} \end{array}$$

	thousands hundreds tens ones				
3.	5456 =	4 th. +	13 h. +	14 t. +	16 o.
	− 2589 =	2 th. +	5 h. +	8 t. +	9 o.
	2867 =	2 th. +	8 h. +	6 t. +	7 o.

Written Exercises

106. Subtract:

1.	2.	3.	4.	5.
163 − 77	386 − 138	291 − 147	975 − 519	500 − 399

6.	7.	8.	9.	10.
842 − 243	191 − 97	308 − 214	501 − 322	328 − 191

11.	12.	13.	14.	15.
300 − 211	400 − 295	500 − 137	600 − 360	700 − 137

16.	17.	18.	19.	20.
3,302 1,103	4,905 2,307	5,873 2,591	4,893 1,506	3,342 2,243

21.	22.	23.	24.	25.
5,561 1,165	4,347 3,748	5,996 3,699	6,327 4,738	8,443 4,378

Written Exercises

107. Find the difference or remainder:

1.	2.	3.	4.	5.
109 − 27	121 − 19	121 − 29	235 − 183	315 − 86

6.	7.	8.	9.	10.
293 − 167	160 − 25	160 − 75	160 − 36	106 − 50

11.	12.	13.	14.	15.
310	314	875	875	578
−210	−125	−169	−309	−390

16. 182 − 52 17. 90 − 64 18. 100 − 26 19. 70 − 18

20. 140 − 35 21. 90 − 48 22. 40 − 28 23. 50 − 24

24. 170 − 54 25. 100 − 85 26. 70 − 45 27. 70 − 31

28. 100 − 36 29. 80 − 32 30. 90 − 32 31. 100 − 42

Check the results by adding the remainder and subtrahend.

The sum should be the minuend.

Subtraction — Money

108. Subtract:

1. $1.72 from $3.65

PROCESS

$3.65
 1.72
$1.93 *Ans.*

EXPLANATION. — Write the numbers, dollars under dollars and cents under cents.

Put the point in the subtrahend under the point in the minuend.

Subtract. Separate the dollars from the cents in the answer by the point, placing it under the point in the minuend and the point in the subtrahend.

When the minuend or the subtrahend does not contain cents, it is advisable, at the beginning, to add the point and the two ciphers.

2.	3.	4.	5.
$13.75	$2.65	$5.95	$10.75
12.25	1.70	3.63	8.65

6.	7.	8.	9.
$ 125.00	$ 235.00	$ 348	$ 97.76
96.00	166.00	179	79.28

10.	11.	12.	13.
$ 65.00	$ 74.23	$ 325.35	$ 818.38
46.38	39.00	138.40	256.96

Written Problems

109. 1. A railroad car will seat 50 persons; 15 seats are vacant. How many seats are occupied?

2. Mary read a fairy tale 19 pages long, another of 15 pages, and a third of 23 pages. How many pages did she read altogether?

3. Julia buys ribbons costing 57 ¢. She gives the clerk 75 ¢. How much change does she receive?

4. There are 52 Sundays in a year of 365 days. How many week days in the year?

5. A pig and a horse cost $ 135. The pig costs $ 13. What is the cost of the horse?

6. John has read 189 pages in a book of 500 pages. How many has he still to read?

7. A newsboy pays $ 3.75 for a hundred papers and sells them for $ 4.90. What is his profit?

8. A boy bought a coat for $ 3.75 and received $ 1.25 change. How much did he give the storekeeper?

9. In Colorado Springs there are about 325 days of sunshine in a year. How many cloudy days in a year of 365 days?

Addition and Subtraction

110. Find the answers:

1. $74 - 28 - 19$?

PROCESS

$$\begin{array}{r} 74 \\ -28 \\ \hline 46 \\ -19 \\ \hline 27 \end{array}$$

EXPLANATION. — Subtract 28 from 74. This gives the remainder 46. Subtract 19 from 46. This gives the remainder 27.

The answer may also be obtained by adding 28 and 19 and subtracting the sum from 74.

2. $125 - 56 + 34$?

PROCESS

$$\begin{array}{r} 125 \\ -56 \\ \hline 69 \\ +34 \\ \hline 103 \end{array}$$

EXPLANATION. — Subtract 56 from 125. This gives the remainder 69. To 69 add 34. This gives the answer, 103.

The answer may also be obtained by adding 125 and 34, and subtracting 56.

Written Exercises

111. Perform the operations indicated:

1. $39 + 44 + 15 = ?$
2. $25 - 12 + 72 = ?$
3. $95 - 59 + 13 = ?$
4. $84 - 48 + 17 = ?$
5. $12 - 7 + 78 = ?$
6. $9 - 6 + 85 = ?$
7. $82 - 36 + 11 = ?$
8. $81 - 41 + 16 = ?$
9. $22 + 24 + 13 = ?$
10. $21 + 13 + 22 = ?$
11. $23 + 12 + 64 = ?$
12. $43 + 13 + 33 = ?$
13. $21 + 43 + 23 = ?$
14. $24 + 52 + 23 = ?$

15. $42 + 31 + 26 = ?$

16. $12 + 34 + 42 = ?$

17. $78 - 23 - 24 = ?$

18. $64 - 23 + 57 = ?$

19. $96 - 43 - 31 = ?$

20. $36 - 14 + 45 = ?$

21. $78 - 33 - 24 = ?$

22. $73 - 41 + 68 = ?$

23. $97 - 32 - 54 = ?$

24. $45 + 32 - 67 = ?$

25. $14 + 84 - 46 = ?$

26. $85 - 64 + 38 = ?$

27. $87 - 34 - 23 = ?$

28. $97 - 72 + 54 = ?$

29. $115 - 61 + 237 = ?$

30. $432 - 271 + 19 = ?$

31. $951 - 891 + 25 = ?$

32. $809 + 100 - 379 = ?$

33. $3 + 309 - 216 = ?$

34. $7 - 3 + 700 - 400 = ?$

35. $\$400 - \$216 - \$109 = ?$

36. $\$11 + \$715 - \$612 = ?$

37. $78¢ - 39¢ + 41¢ = ?$

38. $95¢ \div 79¢ + 38¢ = ?$

MULTIPLICATION

Review

112. 1. How much are 2 twos? 3 twos? 5 twos? 6 twos? 7 twos? 8 twos? 9 twos?

2. How much are 2 nines? 2 sevens? 2 sixes? 2 fours? 2 eights? 2 fives? 2 threes?

3. Give the products:

9	6	8	5	3	1	4
×2	×2	×2	×2	×2	×2	×2

4.

11	12	14	13	10	2	2
×2	×2	×2	×2	×2	×7	×9

5. Make drill device for multiplication by 2.

Written Exercises

113. Find the product:

1. 37 multiplied by 2.

PROCESS

tens ones

$$\begin{array}{r} 37 \\ \times 2 \\ \hline 74 \end{array}$$ *Ans.*

EXPLANATION. — 2 × 7 ones = 14 = 1 ten + 4 ones.
Write the 4.
2 × 3 tens = 6 tens.
6 tens + 1 ten = 7 tens.
Write the 7.
The product is 74.

2.	15	14	18	17	16	20	19 ×2
3.	21	22	24	23	25	26	28 ×2
4.	29	27	30	32	33	31	34 ×2
5.	36	35	37	39	40	41	42 ×2

114. Multiplication by Three

1. How many *A*'s in *B*? How many are 2 threes?

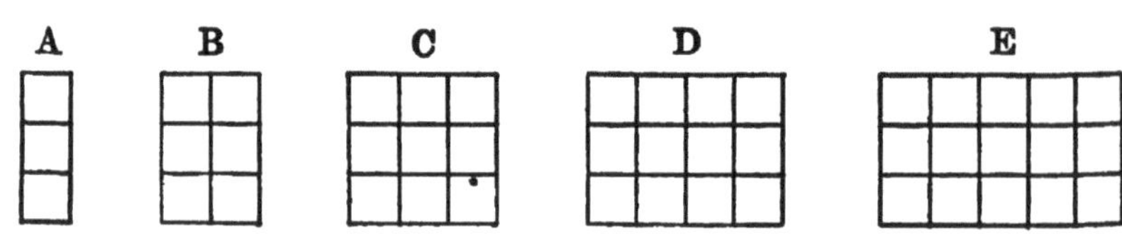

2. How many *A*'s in *C*? How many are 3 threes?

3. How many *A*'s in *D*? How many are 4 threes?

4. How many *A*'s in *E*? How many are 5 threes?

5. Make drawings to show that 6 threes = 18, 7 threes = 21, 8 threes = 24, 9 threes = 27, 10 threes = 30.

115. Memorize the table:

THE MULTIPLICATION TABLE OF 3	
$1 \times 3 = 3$	$6 \times 3 = 18$
$2 \times 3 = 6$	$7 \times 3 = 21$
$3 \times 3 = 9$	$8 \times 3 = 24$
$4 \times 3 = 12$	$9 \times 3 = 27$
$5 \times 3 = 15$	$10 \times 3 = 30$

Make the table of three, beginning with 3×1, 3×2, etc.

Oral Problems

116. 1. There are three feet in one yard. How many feet in 2 yd.? In 4 yd.? In 5 yd.? In 3 yd.? In 6 yd.? In 8 yd.? In 7 yd.? In 9 yd.? In 10 yd.?

2. A man earns $3 a day. How much does he earn in 2 days? In 5 days?

3. How many cents does Frank spend in 3 days, if he spends 3 cents each day?

4. What do you pay for 3 two-cent stamps? For 4 three-cent stamps? For 6 three-cent stamps?

5. What will 3 canary birds cost at $5 apiece?

6. Make and solve a problem about 8 oranges.

7. Make and solve a problem about 3 yd.

8. Make and solve a problem about 5 balls.

117. **Oral Exercises**

1. 2×7	2. 3×2	3. 3×5	4. 9×3	5. 1×8
6. 0×7	7. 8×3	8. 6×2	9. 2×2	10. 1×9
11. 8×0	12. 7×3	13. 3×6	14. 4×3	15. 2×9
16. 3×3	17. 5×2	18. 2×4	19. 0×1	20. 5×1
21. 3×7	22. 8×2	23. 3×8	24. 5×3	25. 3×9

Drill Device — The Wheel

118. Multiply by 3 in one direction, then in the other direction. Change the order of the numbers on the rim of the wheel.

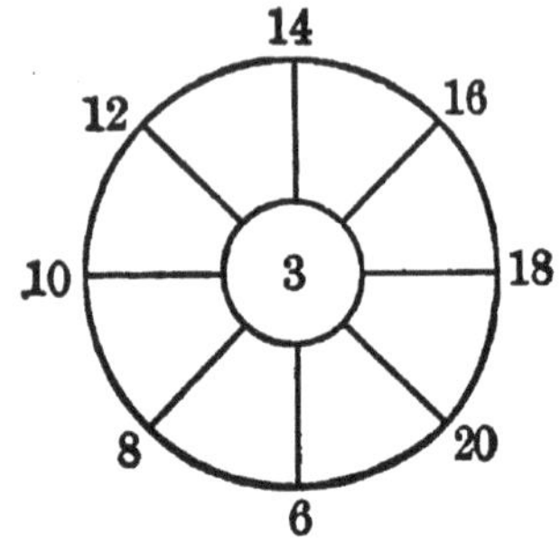

Multiplication

119. Multiply 27 by 3.

Process

$$\begin{array}{r} 27 \\ \times 3 \\ \hline 81 \end{array}$$

Explanation. — 3×7 ones $= 21 = 2$ tens $+ 1$.

Write the 1.

3×2 tens $= 6$ tens.

6 tens $+$ 2 tens $= 8$ tens.

Write the 8.

The product is 81.

Written Exercises

120. Find the products:

1. $\begin{array}{r} 11 \\ \times 3 \\ \hline \end{array}$	2. $\begin{array}{r} 13 \\ \times 3 \\ \hline \end{array}$	3. $\begin{array}{r} 12 \\ \times 3 \\ \hline \end{array}$	4. $\begin{array}{r} 10 \\ \times 3 \\ \hline \end{array}$	5. $\begin{array}{r} 14 \\ \times 3 \\ \hline \end{array}$	6. $\begin{array}{r} 16 \\ \times 3 \\ \hline \end{array}$

7. 20 × 3   8. 17 × 3   9. 18 × 3   10. 19 × 3   11. 21 × 3   12. 25 × 3

13. 24 × 3   14. 27 × 3   15. 23 × 3   16. 22 × 3   17. 29 × 3   18. 30 × 3

121. Written Exercise

> *One dollar and fifty cents* is written $1.50
> *Seventy-five cents* is written $.75
> *Six cents* is written $.06

1. Write nine dollars and thirty cents.
2. Write ten dollars and nine cents.
3. Write twenty-five cents.
4. Ten dollars and one cent.

Multiplication — Money

122. 1. How much are 3 times $1.61?

PROCESS

$1.61
3
$4.83 *Ans.*

EXPLANATION. — Separate the dollars from the cents by the point. The two right-hand digits are cents.

The answer is read: Four dollars *and* eighty-three cents.

2. 3 × $2.15   3. 3 × $2.45   4. 3 × $2.25
5. 3 × $3.50   6. 3 × $1.13   7. 3 × $4.57
8. 3 × $2.96   9. 3 × $.75   10. 3 × $3.05

DIVISION

Review — Division By Two

123. 1. How many 2's in 4? In 6? In 8? In 10?

2. How much is $2\overline{)12}$?
3. How much is $2\overline{)16}$?
4. How much is $2\overline{)18}$?
5. How much is $2\overline{)14}$?
6. Use drill device, Ex. 60, for division.

Division — Without a Remainder

124. Find the quotient:

1. $2\overline{)68}$

PROCESS

$$\begin{array}{r} 34 \\ 2\overline{)68} \end{array}$$

EXPLANATION. — $6 \div 2 = 3$. Write the 3 above the 6.

$8 \div 2 = 4$. Write the 4 above the 8.

The quotient, 34, is the answer.

Check the answer by multiplication, $2 \times 34 = 68$.

2. $2\overline{)24}$	3. $2\overline{)28}$	4. $2\overline{)30}$	5. $2\overline{)26}$
6. $2\overline{)42}$	7. $2\overline{)62}$	8. $2\overline{)40}$	9. $2\overline{)44}$
10. $2\overline{)48}$	11. $2\overline{)66}$	12. $2\overline{)88}$	13. $2\overline{)86}$
14. $2\overline{)84}$	15. $2\overline{)46}$	16. $2\overline{)68}$	17. $2\overline{)80}$

Division by Three — Oral

125. 1. How many 3's in 6?

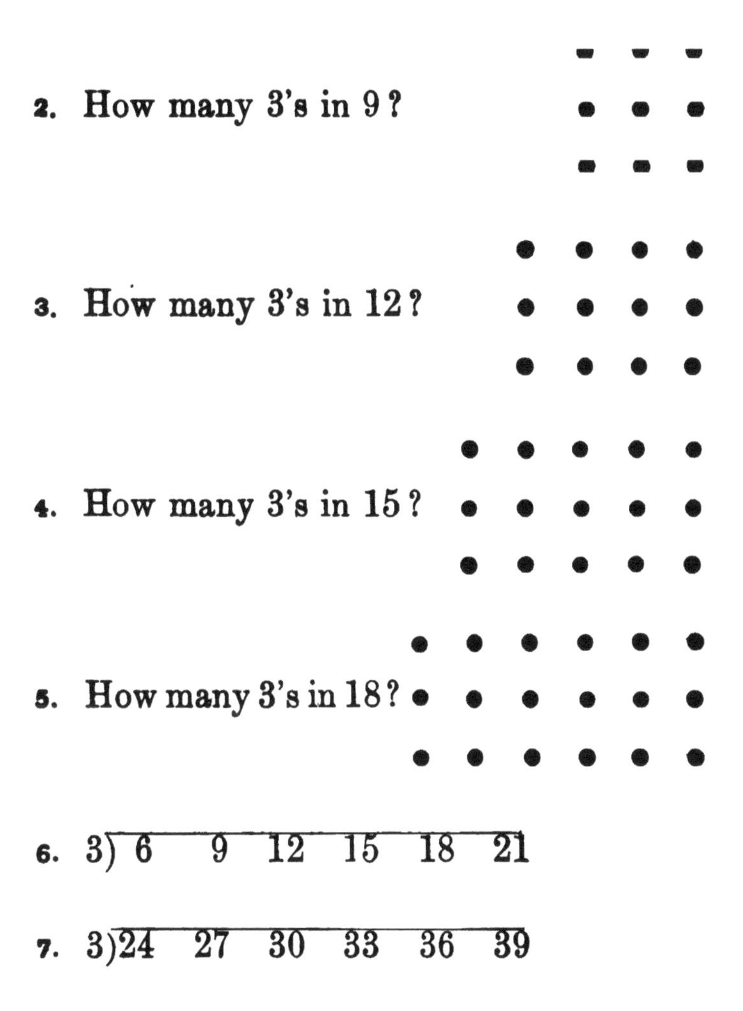

2. How many 3's in 9?

3. How many 3's in 12?

4. How many 3's in 15?

5. How many 3's in 18?

6. 3)‾6   9   12   15   18   21‾

7. 3)‾24   27   30   33   36   39‾

8. Illustrate, by the use of splints, in Exercises 6 and 7.

126. Study the table:

THE DIVISION TABLE OF 3

$3 \div 3 = 1$	$\overset{1}{3\overline{)3}}$	$18 \div 3 = 6$	$\overset{6}{3\overline{)18}}$
$6 \div 3 = 2$	$\overset{2}{3\overline{)6}}$	$21 \div 3 = 7$	$\overset{7}{3\overline{)21}}$
$9 \div 3 = 3$	$\overset{3}{3\overline{)9}}$	$24 \div 3 = 8$	$\overset{8}{3\overline{)24}}$
$12 \div 3 = 4$	$\overset{4}{3\overline{)12}}$	$27 \div 3 = 9$	$\overset{9}{3\overline{)27}}$
$15 \div 3 = 5$	$\overset{5}{3\overline{)15}}$	$30 \div 3 = 10$	$\overset{10}{3\overline{)30}}$

Oral Problems

127. 1. 24 trees are planted in 3 equal rows. How many trees in each row?

2. If 3 boys earn $18 per week, what does each earn?

3. A farmer has 21 bu. of grain which he puts up in bags holding 3 bu. How many bags of grain are there?

4. In a school there are 24 children in 3 classes of equal size. How many children in each class?

5. A storekeeper bought a box of shoes for $ 36. If the shoes cost $ 3 a pair, how many pairs are there in the box?

6. How many yards are there in 12 feet? How many yards are there in 15 feet?

7. How many sets of 3's in 21?

Drill Device — The Wheel

128. Divide the numbers on the rim by the number in the center.

Change the numbers on the rim.

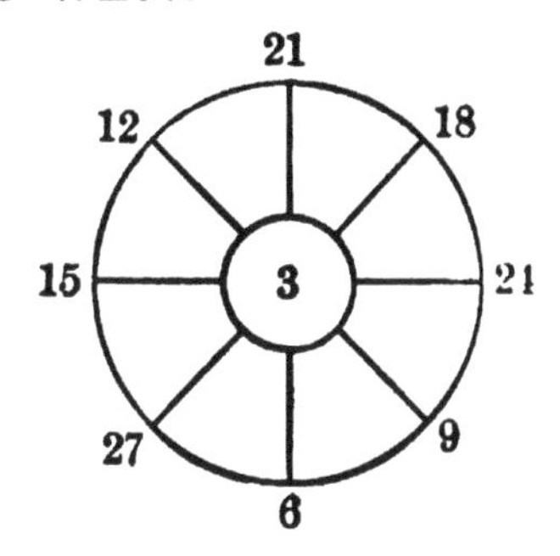

Written Exercises

129. Find the quotient:

1. $3\overline{)39}$

PROCESS

$$\begin{array}{r} 13 \\ 3\overline{)39} \end{array}$$

EXPLANATION. — $3 \div 3 = 1$. Write the 1. $9 \div 3 = 3$. Write the 3. The quotient, 13, is the answer. Check the answer.

2. $3\overline{)33}$	3. $3\overline{)36}$	4. $3\overline{)60}$	5. $3\overline{)66}$	6. $3\overline{)63}$
7. $3\overline{)69}$	8. $3\overline{)90}$	9. $3\overline{)93}$	10. $3\overline{)96}$	11. $3\overline{)99}$

12. $30 \div 3$	13. $39 \div 3$	14. $66 \div 3$
15. $66 \div 3$	16. $333 \div 3$	17. $153 \div 3$

Division by Three

130. 1. Into how many parts is the circle divided?

2. How do these parts compare in size?

3. What is each part called?

4. What are 2 parts called?

5. Draw a line and divide into 3 equal parts.

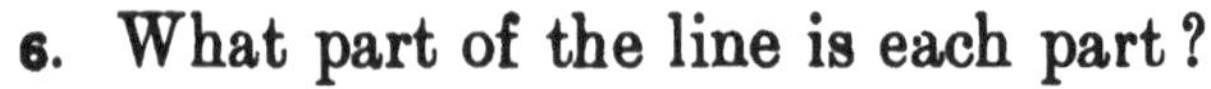

6. What part of the line is each part?

To find **one third** of a number divide the number by **three.**

One third is written $\frac{1}{3}$

Oral Exercises

131. 1. How much is $3\overline{)30}$? How much is $\frac{1}{3}$ of 30?

2. Find $\frac{1}{3}$ of 21, 27, 36, 93, 99.

3. Which is more, $\frac{1}{2}$ of 40 or $\frac{1}{3}$ of 63?

4. Which is more, $\frac{1}{3}$ of 36 or $\frac{1}{3}$ of 33?

5. Find $\frac{1}{3}$ of 15, 30, 45, 24, 60.

Oral Problems

132. 1. How many yards are there in a line 18 ft. long?

2. A boy had 15 marbles. He lost $\frac{1}{3}$ of them. How many did he lose?

3. How many marbles has he now?

4. A dressmaker earns $24 a week. She spends $\frac{1}{3}$ of it. How much has she left?

5. There are 21 words in a spelling lesson. Charles spelled $\frac{1}{3}$ incorrectly. How many did he spell correctly?

Money

133. Divide:

1. $4.84 by 2

PROCESS	EXPLANATION. — After dividing by 2, separate the dollars from the cents by the point.
$2\overline{)\$4.84}$ = $2.42	

2. $3\overline{)\$3.60}$	3. $2\overline{)\$2.48}$	4. $3\overline{)\$6.30}$
5. $3\overline{)\$6.60}$	6. $3\overline{)\$9.06}$	7. $3\overline{)\$6.33}$
8. $3\overline{)\$9.90}$	9. $3\overline{)\$9.33}$	10. $3\overline{)\$9.63}$
11. $3\overline{)\$3.69}$	12. $3\overline{)\$3.03}$	13. $3\overline{)\$3.96}$

Exercises — Measurement

134. 1. Draw lines, 1 in. long, $\frac{1}{2}$ in. long, 2 in. long, 3 in. long.

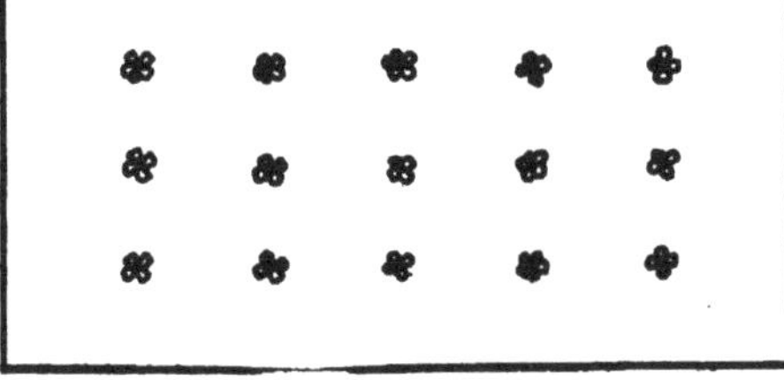

2. Julia has a flower bed 2 yd. long and 1 yd. wide. When telling Mary about it, she makes a drawing, in which 1 *in. stands for* 1 *yd.* Then, 2 inches stand for how many yards?

3. How many yards do 3 in. stand for? $\frac{1}{2}$ in.?

4. In Julia's drawing, 1 in. stands for how many feet? 2 in.? 3 in.?

5. If, on a map, 1 in. stands for 1 mile, how long a line must you draw to stand for 4 miles? $\frac{1}{2}$ mile?

6. If 1 in. stands for 1 mile, what do 2 in. stand for? 3 in.? 5 in.? 8 in.? 9 in.?

7. A certain county is in form of a square, 3 miles on each side. If 1 in. stands for 1 mile, draw a map of the county.

8. Name two places near your home which are one mile apart.

9. About how long does it take you to walk a mile?

10. How many miles can you walk in 2 hours?

Multiplication by Four — Oral

1. How many are $4 + 4$?
2. How many are 2×4?
3. How many are $4 + 4 + 4$?
4. How many are 3×4?
5. How many are $4 + 4 + 4 + 4$?
6. How many are 4×4?
7. How many are $4 + 4 + 4 + 4 + 4$?
8. How many are 5×4?
9. How many are $4 + 4 + 4 + 4 + 4 + 4$?
10. How many are 6×4?

135. Memorize the table:

THE MULTIPLICATION TABLE OF 4	
$1 \times 4 = 4$	$6 \times 4 = 24$
$2 \times 4 = 8$	$7 \times 4 = 28$
$3 \times 4 = 12$	$8 \times 4 = 32$
$4 \times 4 = 16$	$9 \times 4 = 36$
$5 \times 4 = 20$	$10 \times 4 = 40$

Make the table of four, beginning with 4×1, 4×2, etc.

Oral Problems

136. 1. How many leaves are there in 4 three-leaf clovers?

2. How many leaves are there in 5 four-leaf clovers?

3. Aunt Elizabeth has given 4¢ apiece to Ned, Frank, Adam, and Peter. How much has she given in all?

4. Each of four girls gave 7¢ to a beggar. How much did he receive?

5. Martha picked 8 quarts of berries for 4¢ a quart. How much did she earn?

6. If you give 11 marbles to each of 4 boys, how many marbles do you give them altogether?

7. How much must I pay for 4 tables at $10 each?

8. How many shoes does it take to shoe 14 horses?

9. How much will 15 lemons cost at 4¢ apiece?

10. Kate bought 4 yards of cloth at 16¢ a yard. How much did the cloth cost?

11. 18 men buy railroad tickets to a town. If each man pays $4 for his ticket, how much money does the ticket seller receive?

Written Exercises

137. Multiply:

1. 123 × 4	2. 123 × 3	3. 134 × 4	4. 64 × 4	5. 134 × 2	6. 131 × 4
7. 181 × 4	8. 208 × 4	9. 511 × 4	10. 412 × 4	11. 142 × 4	12. 204 × 4
13. 304 × 4	14. 143 × 2	15. 143 × 3	16. 444 × 4	17. 333 × 3	18. 222 × 4

19. 121 × 4	20. 213 × 4	21. 413 × 4	22. 214 × 4
23. 413 × 2	24. 132 × 4	25. 124 × 4	26. 124 × 3
27. 124 × 2	28. 104 × 4	29. 24 × 4	30. 94 × 4

Written Problems

138. 1. A clothier sells 4 overcoats at $20 each. How much does he receive for the overcoats?

2. If Mr. Smith saves $100 a year, how much does he save in 4 years?

3. Four large elephant tusks weigh on an average 114 pounds each. What is their combined weight?

4. At $84 apiece, what is the cost of 4 cows?

5. A boy sold 44 newspapers a day for 4 days. How many papers did he sell?

6. A butcher buys 4 turkeys, each weighing 18 lbs. If he pays 30¢ a pound for them, how much do they cost?

7. Steak is worth 36¢ a pound. How much will 4 lb. cost?

8. A merchant sold 48 pairs of shoes at $4 a pair. How much did he receive for the shoes?

9. A farmer has 4 acres of rye. On each acre there are 24 bu. How many bushels on the 4 acres?

10. A real estate dealer sold 4 lots at $135 each. How much did he receive for them?

Division by Four — Oral

139. 1. How many 4's in 8?

2. How many 4's in 12?

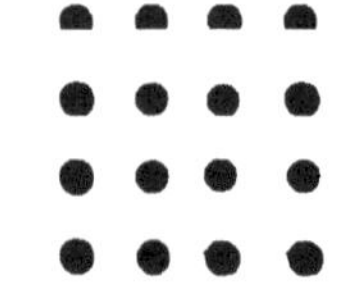

3. How many 4's in 16?

4. How many 4's in 20?

5. How many 4's in 24?

140. Study the table:

THE DIVISION TABLE OF 4

$4 \div 4 = 1$	$\begin{array}{r}1\\4\overline{)4}\end{array}$	$24 \div 4 = 6$	$\begin{array}{r}6\\4\overline{)24}\end{array}$
$8 \div 4 = 2$	$\begin{array}{r}2\\4\overline{)8}\end{array}$	$28 \div 4 = 7$	$\begin{array}{r}7\\4\overline{)28}\end{array}$
$12 \div 4 = 3$	$\begin{array}{r}3\\4\overline{)12}\end{array}$	$32 \div 4 = 8$	$\begin{array}{r}8\\4\overline{)32}\end{array}$
$16 \div 4 = 4$	$\begin{array}{r}4\\4\overline{)16}\end{array}$	$36 \div 4 = 9$	$\begin{array}{r}9\\4\overline{)36}\end{array}$
$20 \div 4 = 5$	$\begin{array}{r}5\\4\overline{)20}\end{array}$	$40 \div 4 = 10$	$\begin{array}{r}10\\4\overline{)40}\end{array}$

Oral Problems

141. 1. In one gallon there are 4 quarts; how many gallons are there in 8 quarts? In 16 quarts?

2. 12 apples are to be divided equally among 4 children. How many does each child get?

3. At 4¢ a pound, how many pounds of sugar can you buy with 40¢?

4. A farmer separated his flock of 28 sheep into 4 equal groups. How many sheep in each group?

5. A man divided $32 equally among 4 boys. How much did each receive?

Division by Four

142. 1. Into how many parts is the square divided?

2. How do these parts compare in size?

3. What part of the whole is each part?

4. What is each part called?

5. What are 2 parts called?

6. What are 3 parts called?

7. Draw a circle and divide it into 4 equal parts.

8. What part of the circle is each part?

To find **one fourth** of a number divide the number by **four.**

One fourth is written $\frac{1}{4}$

Oral Exercises

143.

1. $4\overline{)8}=?$
2. $\frac{1}{4}$ of $8=?$
3. $4\overline{)16}=?$
4. $\frac{1}{4}$ of $16=?$
5. $4\overline{)12}=?$
6. $\frac{1}{4}$ of $12=?$
7. $4\overline{)20}=?$
8. $\frac{1}{4}$ of $20=?$
9. $4\overline{)28}=?$
10. $\frac{1}{4}$ of $28=?$
11. $4\overline{)32}=?$
12. $\frac{1}{4}$ of $32=?$
13. $4\overline{)24}=?$
14. $\frac{1}{4}$ of $24=?$
15. $4\overline{)36}=?$
16. $\frac{1}{4}$ of $36=?$
17. $\frac{1}{4}$ of $40=?$
18. $4\overline{)40}=?$

Drill Device — The Wheel

144. Divide each number on the rim by the number in the center.

Change the numbers on the rim.

Change the order of the numbers.

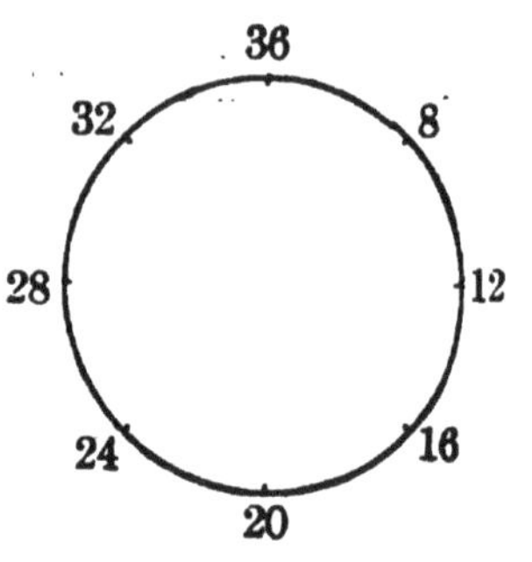

Written Exercises

145. Divide:

1. 124 by 4.

PROCESS

$$\begin{array}{r} 31 \\ 4\overline{)124} \end{array}$$

EXPLANATION

$$\begin{aligned} 4\overline{)124} &= 4\overline{)1 \text{ hundred} + 2 \text{ tens} + 4 \text{ ones}} \\ &= 4\overline{)0 \text{ hundred} + 12 \text{ tens} + 4 \text{ ones}} \\ &= 4\overline{)\qquad\qquad 12 \text{ tens} + 4 \text{ ones}} \\ &\qquad\qquad\qquad\quad 3 \text{ tens} + 1 \text{ one} \\ &= 4\overline{)\qquad\qquad 12 \text{ tens} + 4 \text{ ones}} \\ &\quad\ \ 31 \\ &= 4\overline{)124} \end{aligned}$$

31 is the quotient.

2. $4\overline{)128}$	3. $4\overline{)88}$	4. $4\overline{)84}$	5. $4\overline{)164}$	6. $4\overline{)204}$
7. $4\overline{)148}$	8. $4\overline{)244}$	9. $4\overline{)248}$	10. $4\overline{)280}$	11. $4\overline{)284}$
12. $4\overline{)844}$	13. $4\overline{)484}$	14. $4\overline{)448}$	15. $4\overline{)804}$	16. $3\overline{)123}$
17. $3\overline{)129}$	18. $2\overline{)128}$	19. $3\overline{)126}$	20. $3\overline{)150}$	21. $3\overline{)159}$
22. $3\overline{)210}$	23. $2\overline{)220}$	24. $2\overline{)468}$	25. $3\overline{)690}$	26. $4\overline{)884}$

Written Problems

146. 1. Divide 64¢ equally among 4 boys.

2. Divide 48 bu. of oats into 4 equal parts.

3. The distance from A to B is 36 mi. Find one third of the distance.

4. Draw a line to represent the distance from A to B in problem 3. Mark off one third of the distance.

5. Four masons build the walls of a house and are paid $ 124. What is the share of each?

6. A four-story building is 48 ft. high. How high is each story?

7. How many 4-cent stamps can be bought with 84¢?

8. A merchant pays $ 240 for rent in 4 months. How much does he pay for rent each month?

9. A storekeeper arranged 320 apples in 4 equal piles. How many apples in each pile?

10. An airship travels 244 mi. in 4 hr. How

Measurement — Exercises

147. 1. How do the sides of a square compare with each other in length? If one side measures 10 ft., what is the distance around the square?

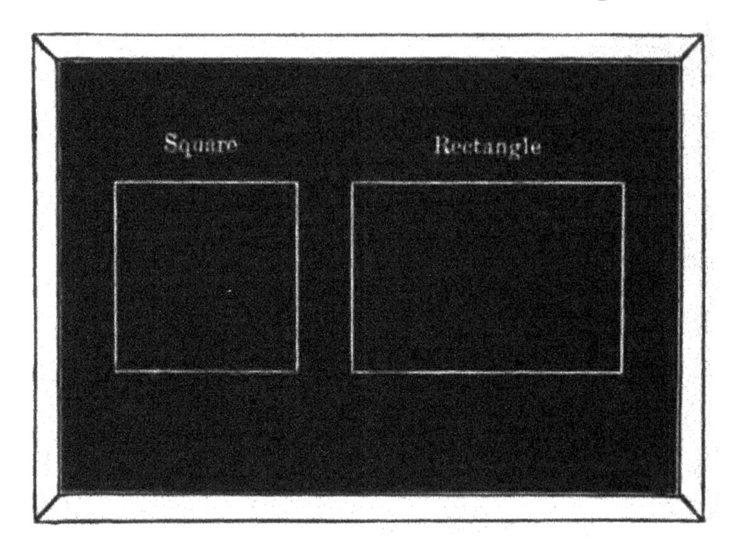

2. It takes 24 minutes to walk around a park in the form of a square. How long will it take to walk the length of one side?

3. How many times longer than wide is the rectangle in our drawing? If its width stands for 10 feet, what does its length stand for? What do all four sides together stand for?

4. How many such squares does it take to exactly cover that rectangle? How many times larger is the rectangle than the square?

5. Draw a square. Then draw a rectangle of the same height as this square, but three times as long.

6. If the width of the rectangle just drawn is 4 miles, what is the length? What is the sum of the four sides?

7. How many squares like the one last drawn will exactly cover the rectangle? How many times larger is this rectangle than the square?

REVIEW

148. Add at sight:

1.	10	11	21	31	41	51	61	71	81	91
	10	10	10	10	10	10	10	10	10	10
2.	10	11	21	31	41	51	61	71	81	91
	20	20	20	20	20	20	20	20	20	20
3.	10	11	21	31	41	51	61	71	81	91
	30	30	30	30	30	30	30	30	30	30
4.	10	11	21	31	41	51	61	71	81	91
	40	40	40	40	40	40	40	40	40	40

5. Add 50, 60, 70, etc., in the same way.

6.	7.	8.	9.	10.
3	4	9	4	7
5	6	6	7	5
7	8	3	2	3
9	2	6	8	9
2	3	4	5	2

149. Subtract at sight:

1.	13	23	33	43	53	63	73	83	93
	4	4	4	4	4	4	4	4	4
2.	14	24	34	44	54	64	74	84	94
	5	5	5	5	5	5	5	5	5
3.	15	25	35	45	55	65	75	85	95
	6	6	6	6	6	6	6	6	6
4.	16	26	36	46	56	66	76	86	96
	7	7	7	7	7	7	7	7	7

5. Subtract 8, 9 in the same way, but change the one's figure in the minuend.

6.	7.	8.	9.	10.
24	33	45	22	31
−7	−8	−6	−9	−5

150. Multiply at sight:

1.	2	3	4	5	6	7	8	9
								×2
2.	2	3	4	5	6	7	8	9
								×3
3.	2	3	4	5	6	7	8	9
								×4

4.	11	21	31	41	51	61	71
							×2

5. 22 32 42 52 62 72 ×3

6. 41 51 61 71 81 91 ×4

7. Multiply 22, 31, 42, 51, etc., by 2, 3, 4.

151. Divide at sight:

1. 2)12 22 32 42 52 62

2. 3)12 21 33 24 18 30

3. 4) 8 24 36 28 40 48

152. Give the answers at sight:

1. $\frac{1}{2}$ of 4, 8, 10, 12, 16.

2. $\frac{1}{3}$ of 6, 9, 12, 15, 18.

3. $\frac{1}{4}$ of 4, 8, 12, 16, 24.

Multiplication by Five — Oral

153. 1. How many squares high is the figure?

2. How many squares in the first row or column?

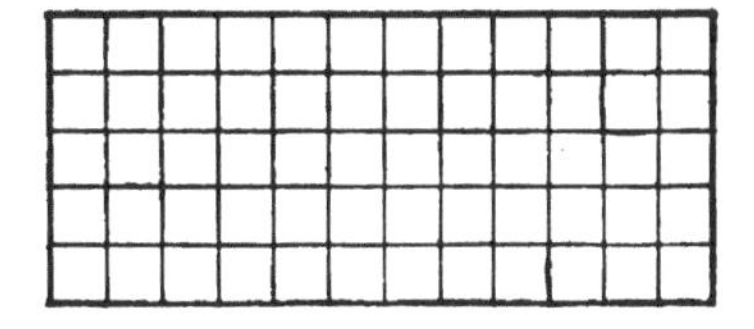

3. How many in the second?

4. How many in the first two?

5. How many in the first three?

6. How many are 1×5? 2×5?

7. How many are 3×5? 4×5?

8. How many are 5×5? 6×5? etc.

154. Memorize the table:

THE MULTIPLICATION TABLE OF 5	
$1 \times 5 = 5$	$6 \times 5 = 30$
$2 \times 5 = 10$	$7 \times 5 = 35$
$3 \times 5 = 15$	$8 \times 5 = 40$
$4 \times 5 = 20$	$9 \times 5 = 45$
$5 \times 5 = 25$	$10 \times 5 = 50$

Make the table of five, beginning with 5×1, 5×2, etc.

Oral Problems

155. 1. A seven-story building has 5 windows in each story. How many windows in the building?

2. How many pupils are there in 5 rows, 8 pupils in each row?

3. How many cents are equal to 6 five-cent pieces?

4. How many cents in 10 five-cent pieces?

5. How many cents in 5 ten-cent pieces?

6. How many dollars in 5 two-dollar bills?

7. How many dollars in 2 five-dollar bills?

8. Five boys visited a public park. The fare each way is 5¢. How much car fare did the boys pay?

9. Fanny paid for her lunch and that of her 4 schoolmates. If each lunch cost 9¢, how much did she pay?

10. How much greater is 8×5 than 4×5? Can you explain this by drawings?

Oral Exercises

156. Give the answers rapidly:

1. 5×9	2. 9×5	3. 5×1	4. 1×5
5. 2×5	6. 5×2	7. 5×3	8. 3×5
9. 5×10	10. 10×5	11. 5×11	12. 11×5
13. 5×4	14. 4×5	15. 7×5	16. 8×5
17. 5×7	18. 5×8	19. 5×5	20. 5×6

Drill Device

157. 1. 4 7 9 10 8 6 3 5 2
×5

2. $9 $6 $5 $4 $7 $8 $3
× 5

3. Multiply each of the following by 5:

4 lb., 7 in., 9 ft., 8 qt., 3 yd., 6 mi.

Written Problems

158 1. If one barrel of sugar cost $18, how much will 5 barrels cost?

2. John owes 13 boys a nickel each. How much does he owe altogether?

3. A boy at a soda fountain sold 23 5-cent soda tickets. How much money did he take in?

4. If there are 5 radishes in a bunch, how many radishes in 27 bunches?

5. There are 5 desks in a row. How many desks in 16 rows?

6. If 3 lemons cost 5¢, how much would 12 lemons cost? (*Compare* 3 *and* 12.)

7. If a half-pint jar of jelly costs 5¢, how much will 2 qt. cost?

8. A family spends 5¢ a day for ice. How much does the ice cost them for a week?

9. A family buys 5 qt. of milk a day at 9¢ a quart. How much does the milk cost for one day? For one week?

Written Exercises

159. Multiply:

1.	2.	3.	4.	5.	6.
125	467	784	603	502	615
×5	×5	×5	×5	×5	×5

7.	8.	9.	10.	11.	12.
933	807	735	666	743	487
×5	×5	×5	×5	×5	×5

13.	36×5	14.	79×5	15.	33×5
16.	21×5	17.	83×5	18.	75×5
19.	97×5	20.	85×5	21.	81×5
22.	67×5	23.	101×5	24.	215×5
25.	695×5	26.	587×5	27.	324×5

28. 55 by 5, by 4, by 3, by 2.

29. 35 by 5, by 3, by 4, by 2.

30. 53 by 2, by 3, by 4, by 5.

Division by Five — Oral

160. 1. How many fives in $5\overline{)10}$ is?

2. How much is $5\overline{)20}$? $5\overline{)15}$? $5\overline{)30}$? $5\overline{)35}$? $5\overline{)25}$? $5\overline{)45}$? $5\overline{)40}$?

3. What is $\frac{1}{5}$ of 20? $\frac{1}{5}$ of 15? $\frac{1}{5}$ of 30?

4. Lucy is paid $\frac{1}{2}$ dollar in dimes? How many coins does she get?

5. How many nickels are worth a quarter of a dollar?

6. How many packages of torpedoes at 5¢ a package can you buy for 30¢? For 40¢? For 45¢?

To find **one fifth** of a number divide the number by **five**.

161. Study the table:

THE DIVISION TABLE OF 5

$5 \div 5 = 1$	$\begin{array}{r}1\\5\overline{)5}\end{array}$	$30 \div 5 = 6$	$\begin{array}{r}6\\5\overline{)30}\end{array}$
$10 \div 5 = 2$	$\begin{array}{r}2\\5\overline{)10}\end{array}$	$35 \div 5 = 7$	$\begin{array}{r}7\\5\overline{)35}\end{array}$
$15 \div 5 = 3$	$\begin{array}{r}3\\5\overline{)15}\end{array}$	$40 \div 5 = 8$	$\begin{array}{r}8\\5\overline{)40}\end{array}$
$20 \div 5 = 4$	$\begin{array}{r}4\\5\overline{)20}\end{array}$	$45 \div 5 = 9$	$\begin{array}{r}9\\5\overline{)45}\end{array}$
$25 \div 5 = 5$	$\begin{array}{r}5\\5\overline{)25}\end{array}$	$50 \div 5 = 10$	$\begin{array}{r}10\\5\overline{)50}\end{array}$

Drill Device — The Wheel

162. Divide each number on the rim by the number in the center.

Place new numbers that are exactly divisible by 5 on the rim.

Drill to find one fifth of a number.

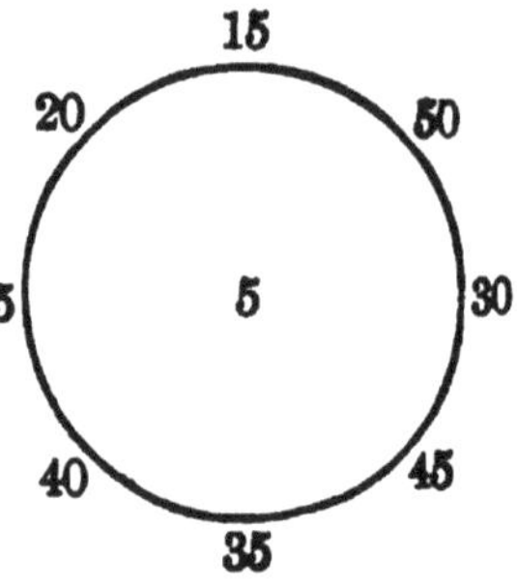

Oral Exercises

163. Give the answers:

1. $5\overline{)5 \quad 15 \quad 35 \quad 50 \quad 10}$

2. $5\overline{)25 \quad 40 \quad 45 \quad 30 \quad 20}$

3. $? \times 5 = 35$ 4. $? \times 5 = 40$ 5. $? \times 5 = 50$

6. $? \times 5 = 45$ 7. $? \times 5 = 55$ 8. $? \times 5 = 25$

9. $7 \times ? = 35$ 10. $8 \times ? = 40$ 11. $10 \times ? = 50$

12. $\frac{1}{5}$ of 20 13. $\frac{1}{5}$ of 45 14. $\frac{1}{5}$ of 35

Written Exercises

164. Find the answers:

1. $5\overline{)105}$ 2. $5\overline{)255}$ 3. $5\overline{)305}$ 4. $5\overline{)355}$

5. $5\overline{)150}$ 6. $5\overline{)155}$ 7. $5\overline{)205}$ 8. $5\overline{)400}$

9. How much is $\frac{1}{5}$ of 55?

10. Take $\frac{1}{5}$ of 50, 100, 25, 250, 500.

11. $455 \div 5$ 12. $450 \div 5$ 13. $505 \div 5$ 14. $555 \div 5$

15. $550 \div 5$ 16. $350 \div 5$ 17. $165 \div 5$ 18. $200 \div 5$

Written Problems

165. 1. For 55¢ I can get —— oranges at 5¢ each.

2. A boy puts 55 beets into bunches of 5 beets each. How many bunches do they make?

3. Draw a line 5 inches long. If 1 inch stands for 12 ft., what does this line stand for?

4. On this line draw a rectangle wide enough to represent a flower bed 12 ft. wide and 60 ft. long. Into how many square patches, 12 ft. on a side, can this flower bed be divided? Complete the drawing so as to show these patches.

5. Draw a rectangle 2 inches long and 1 inch wide. If 1 inch stands for 5 yards, into how many squares, five yards on a side, can this flower bed be divided?

6. At $5 each, how many sheep can you buy for $105?

7. If 5 boys earn $25 a month, 1 boy earns $\frac{1}{5}$ of $25, or $? a month.

8. A coal dealer sold 525 tons of coal in 5 months. What was the average amount sold each month?

9. A railroad train travels 175 mi. in 5 hr. What is the average distance traveled each hour?

10. A man rows a boat 15 mi. He rows at the rate of 5 mi. an hour. How many hours does he row?

Multiplication by Six — Oral

166. 1. Count by 2's to 20.

2. Count by 3's to 30.

3. Count by 6's to 60.

4. What is the sum of five 6's? 5×6?

5. What is the sum of six 6's? 6×6?

6. What is the sum of seven 6's? 7×6?

7. What is the sum of eight 6's, nine 6's, ten 6's?

8. How much is 8×6? 9×6? 10×6?

167. Memorize the table:

THE MULTIPLICATION TABLE OF 6	
$1 \times 6 = 6$	$6 \times 6 = 36$
$2 \times 6 = 12$	$7 \times 6 = 42$
$3 \times 6 = 18$	$8 \times 6 = 48$
$4 \times 6 = 24$	$9 \times 6 = 54$
$5 \times 6 = 30$	$10 \times 6 = 60$

Make the table of six, beginning with 6×1, 6×2, etc.

Division by Six — Oral

168. 1. How much are 6×1? $6 \div 6 = ?$

2. How much are 6×2? $12 \div 6 = ?$

3. How much are 6×3? $18 \div 6 = ?$

4. How much are 6×4? $24 \div 6 = ?$

5. How much are 6×5? $30 \div 6 = ?$

6. What is $\frac{1}{6}$ of 6? $\frac{1}{6}$ of 12?

7. What is $\frac{1}{6}$ of 18? $\frac{1}{6}$ of 24? $\frac{1}{6}$ of 30?

8. $6\overline{)6 \quad 12 \quad 18 \quad 24 \quad 30}$

9. $6\overline{)36 \quad 42 \quad 48 \quad 54 \quad 60}$

10. Illustrate Exercises 1, 2, 3 by splints.

To find **one sixth** of a number divide the number by **six**.

169. Study the table:

THE DIVISION TABLE OF 6

$6 \div 6 = 1$	$6\overline{)6}$ = 1	$36 \div 6 = 6$	$6\overline{)36}$ = 6
$12 \div 6 = 2$	$6\overline{)12}$ = 2	$42 \div 6 = 7$	$6\overline{)42}$ = 7
$18 \div 6 = 3$	$6\overline{)18}$ = 3	$48 \div 6 = 8$	$6\overline{)48}$ = 8
$24 \div 6 = 4$	$6\overline{)24}$ = 4	$54 \div 6 = 9$	$6\overline{)54}$ = 9
$30 \div 6 = 5$	$6\overline{)30}$ = 5	$60 \div 6 = 10$	$6\overline{)60}$ = 10

Drill Device — The Circle

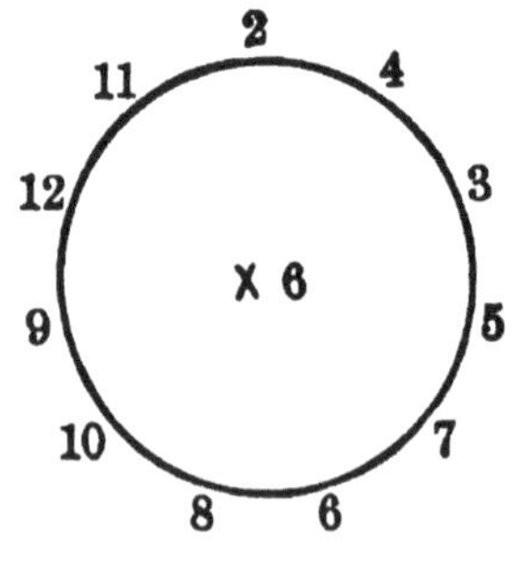

170. Multiply the numbers on the circle by the number in the center. Change the order of the numbers on the circle.

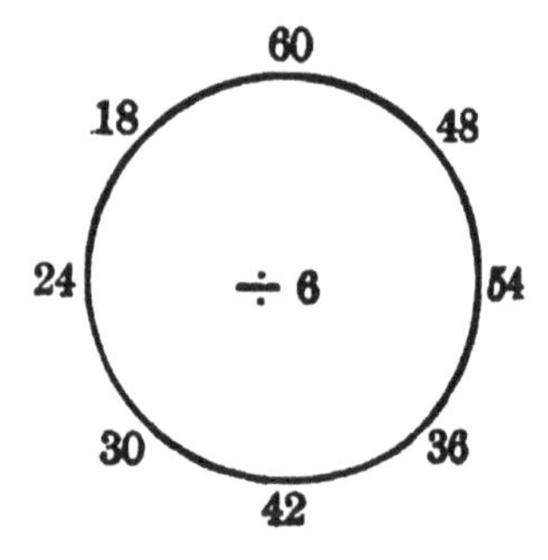

Divide the numbers on the circle by the number in the center. Change the order of the numbers. Change the numbers.

Multiplication — Written Exercises

171. Find the answers:

1. $\begin{array}{r} 10 \\ \times 6 \\ \hline \end{array}$ 2. $\begin{array}{r} 13 \\ \times 6 \\ \hline \end{array}$ 3. $\begin{array}{r} 15 \\ \times 6 \\ \hline \end{array}$ 4. $\begin{array}{r} 14 \\ \times 6 \\ \hline \end{array}$ 5. $\begin{array}{r} 12 \\ \times 6 \\ \hline \end{array}$

6. $\begin{array}{r} 11 \\ \times 6 \\ \hline \end{array}$ 7. $\begin{array}{r} 16 \\ \times 6 \\ \hline \end{array}$ 8. $\begin{array}{r} 18 \\ \times 6 \\ \hline \end{array}$ 9. $\begin{array}{r} 17 \\ \times 6 \\ \hline \end{array}$ 10. $\begin{array}{r} 19 \\ \times 6 \\ \hline \end{array}$

11. $\begin{array}{r} 20 \\ \times 6 \\ \hline \end{array}$ 12. $\begin{array}{r} 22 \\ \times 6 \\ \hline \end{array}$ 13. $\begin{array}{r} 25 \\ \times 6 \\ \hline \end{array}$ 14. $\begin{array}{r} 26 \\ \times 6 \\ \hline \end{array}$ 15. $\begin{array}{r} 31 \\ \times 6 \\ \hline \end{array}$

16. 111×6 17. 212×6 18. 95×6 19. 73×6

20. 47×6 21. 83×6 22. 59×6 23. 99×6

24. 78×6 25. 101×6 26. 6×39 27. 6×191

28. 55×6 29. 6×66 30. 6×77 31. 6×109

Division — Written Exercises

172. Find the answers:

1. $\frac{1}{6}$ of 66 2. $\frac{1}{6}$ of 84 3. $\frac{1}{6}$ of 108

4. $\frac{1}{6}$ of 96 5. $\frac{1}{6}$ of 126 6. $\frac{1}{6}$ of 114

7. $6\overline{)66}$ 8. $6\overline{)606}$ 9. $6\overline{)660}$ 10. $6\overline{)126}$

11. $6\overline{)180}$ 12. $6\overline{)186}$ 13. $6\overline{)246}$ 14. $6\overline{)306}$

15. $6\overline{)300}$ 16. $6\overline{)426}$ 17. $6\overline{)480}$ 18. $6\overline{)546}$

19. $6\overline{)540}$ 20. $6\overline{)600}$ 21. $6\overline{)606}$ 22. $6\overline{)360}$

Written Problems

173. 1. In 6 months John works 120 days. In 1 month he works $\frac{1}{6}$ of 120 days, or —— days.

2. How many ink erasers at 6¢ apiece can be purchased for 66¢?

3. Six windows contain 36 panes of glass. How many panes are there in each window?

4. A street car runs 660 yards in 6 minutes, or —— yards in 1 minute.

5. In 1 month there are 20 days of school. How many days in 6 months?

6. A man works 24 days each month. In 6 months he works —— days.

7. A man works 126 days in 6 months. In one month he worked $\frac{1}{6}$ of 126 days, or —— days.

Review — Tables

174. Study the tables.

Table of Two	Table of Three	Table of Four	Table of Five	Table of Six
$1 \times 2 = 2$	$1 \times 3 = 3$	$1 \times 4 = 4$	$1 \times 5 = 5$	$1 \times 6 = 6$
$2 \times 2 = 4$	$2 \times 3 = 6$	$2 \times 4 = 8$	$2 \times 5 = 10$	$2 \times 6 = 12$
$3 \times 2 = 6$	$3 \times 3 = 9$	$3 \times 4 = 12$	$3 \times 5 = 15$	$3 \times 6 = 18$
$4 \times 2 = 8$	$4 \times 3 = 12$	$4 \times 4 = 16$	$4 \times 5 = 20$	$4 \times 6 = 24$
$5 \times 2 = 10$	$5 \times 3 = 15$	$5 \times 4 = 20$	$5 \times 5 = 25$	$5 \times 6 = 30$
$6 \times 2 = 12$	$6 \times 3 = 18$	$6 \times 4 = 24$	$6 \times 5 = 30$	$6 \times 6 = 36$
$7 \times 2 = 14$	$7 \times 3 = 21$	$7 \times 4 = 28$	$7 \times 5 = 35$	$7 \times 6 = 42$
$8 \times 2 = 16$	$8 \times 3 = 24$	$8 \times 4 = 32$	$8 \times 5 = 40$	$8 \times 6 = 48$
$9 \times 2 = 18$	$9 \times 3 = 27$	$9 \times 4 = 36$	$9 \times 5 = 45$	$9 \times 6 = 54$
$10 \times 2 = 20$	$10 \times 3 = 30$	$10 \times 4 = 40$	$10 \times 5 = 50$	$10 \times 6 = 60$

Multiplication by Seven — Oral

175. Give the answers:

1. Add two 7's. $2 \times 7 = ?$
2. Add three 7's. $3 \times 7 = ?$
3. Add four 7's. $4 \times 7 = ?$
4. Add five 7's. $5 \times 7 = ?$
5. In a window there are 8 panes of glass. How many panes are there in 7 windows?
6. A man bought 5 lb. of sugar at 7¢ a pound. How much did he pay?

176. Memorize the table:

THE MULTIPLICATION TABLE OF 7	
$1 \times 7 = 7$	$6 \times 7 = 42$
$2 \times 7 = 14$	$7 \times 7 = 49$
$3 \times 7 = 21$	$8 \times 7 = 56$
$4 \times 7 = 28$	$9 \times 7 = 63$
$5 \times 7 = 35$	$10 \times 7 = 70$

Make the table of seven, beginning with 7×1, 7×2, etc.

Division by Seven — Oral

177. Give the answers:

1. How many 7's in 21? $21 \div 7 = ?$
2. How many 7's in 28? $28 \div 7 = ?$
3. How many 7's in 35? $35 \div 7 = ?$
4. How many 7's in 42? $42 \div 7 = ?$

5. What is $\frac{1}{7}$ of 14? Of 21?

6. What is $\frac{1}{7}$ of 28? Of 35?

7. $7\overline{)7 \quad 14 \quad 21 \quad 28 \quad 35}$

To find **one seventh** of a number divide the **number** by **seven.**

178. Study the table:

THE DIVISION TABLE OF 7

$7 \div 7 = 1$	$7\overline{)7}$ with quotient 1	$42 \div 7 = 6$	$7\overline{)42}$ with quotient 6
$14 \div 7 = 2$	$7\overline{)14}$ with quotient 2	$49 \div 7 = 7$	$7\overline{)49}$ with quotient 7
$21 \div 7 = 3$	$7\overline{)21}$ with quotient 3	$56 \div 7 = 8$	$7\overline{)56}$ with quotient 8
$28 \div 7 = 4$	$7\overline{)28}$ with quotient 4	$63 \div 7 = 9$	$7\overline{)63}$ with quotient 9
$35 \div 7 = 5$	$7\overline{)35}$ with quotient 5	$70 \div 7 = 10$	$7\overline{)70}$ with quotient 10

Drill Device — The Circle

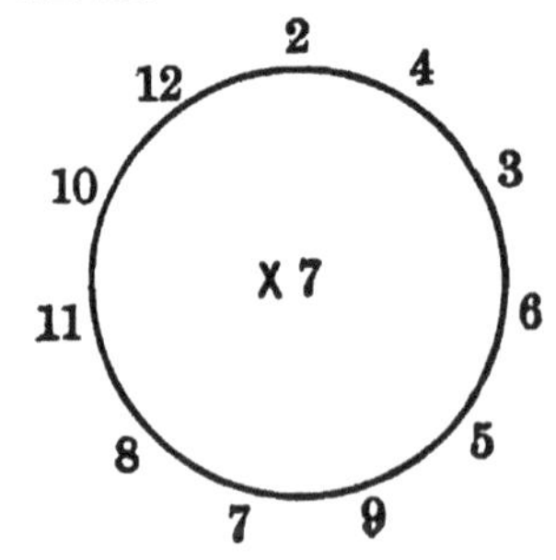

179. Multiply the numbers on the circle by the number in the center.

Change the order of the numbers on the circle.

Divide the numbers on the circle by the number in the center.

Change the order of the numbers on the circle.

Change the numbers on the circle.

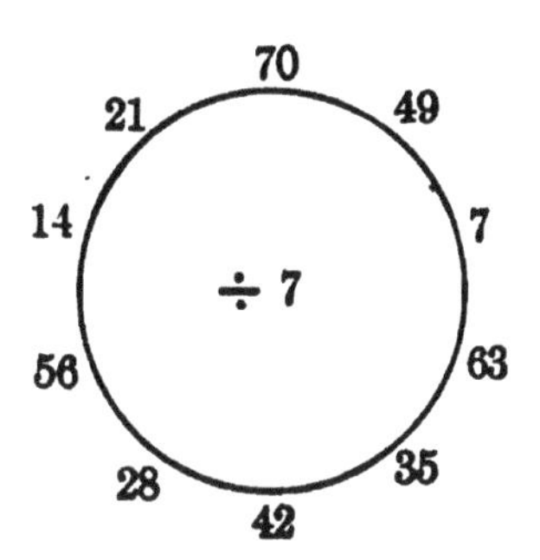

Multiplication — Written Exercises

180. Find the answers:

1. 15 × 7	2. 13 × 7	3. 12 × 7	4. 14 × 7	5. 11 × 7	6. 17 × 7
7. 21 × 7	8. 20 × 7	9. 23 × 7	10. 19 × 7	11. 25 × 7	12. 30 × 7

13. 111 123 93 89 67 109 190 × 7

14. 43 56 78 91 82 64 100 × 7

15. $60 \times 2 \times 7$

16. $30 \times 7 \times 3$

17. $20 \times 3 \times 7$

18. $11 \times 4 \times 7$

Division — Written Exercises

181. Find the answers:

1. $\frac{1}{7}$ of 49
2. $\frac{1}{7}$ of 28
3. $\frac{1}{7}$ of 84
4. $\frac{1}{7}$ of 70
5. $\frac{1}{7}$ of 35
6. $\frac{1}{7}$ of 56

7. $7\overline{)77}$	8. $7\overline{)140}$	9. $7\overline{)147}$	10. $7\overline{)217}$
11. $7\overline{)280}$	12. $7\overline{)210}$	13. $7\overline{)287}$	14. $7\overline{)350}$
15. $7\overline{)427}$	16. $7\overline{)357}$	17. $7\overline{)497}$	18. $7\overline{)490}$

Multiplication — Oral Problems

182. 1. How many days in 8 wk.? In 7 wk.?

2. How many days in 2 wk. of vacation?

3. A fire ladder is made of 7 parts. Each part is 5 ft. long. How long is the ladder?

4. What is the cost of 7 notebooks at 5¢ each?

5. A workman receives $2 a day. How much does he earn in 5 da.?

6. How much greater is 7×8 than 7×4?

7. A boy can row a boat 3 mi. an hour. How far can he row in 7 hr. at the same rate?

8. A merchant sold 7 pairs of shoes at $4 a pair. How much did he receive for the shoes?

Division — Oral Problems

183. 1. A fruit dealer arranged 35 apples in 5 piles. How many apples in each pile?

2. A line 84 in. long is divided into 7 equal parts. How many inches in each part?

3. How many weeks in 77 days?

4. How much greater is $35 \div 7$ than $14 \div 7$?

5. Mr. Hall has 56 hills of corn arranged in 7 rows. How many hills in each row?

6. John has 77 marbles. His brother has $\frac{1}{7}$ as many. How many has his brother?

7. If 7 oranges cost 28¢, how much would 1 orange cost?

8. How many balls in $\frac{1}{7}$ of 21 balls?

9. How many cents in $\frac{1}{7}$ of 49¢?

10. If 7 bottles of ink cost 35¢, what is the price of 1 bottle?

Written Problems

184. 1. How many weeks are there is 357 days?

2. If there are 30 days in a month, how many days are there in 7 months?

3. William's mother gave him 63¢, and told him to buy berries. If the berries cost 7¢ a box, how many boxes did William buy?

4. If a stage runs 7 miles an hour, how far will it run in 13 hours?

5. Father pays $28 for coal at $7 a ton. How many tons does he purchase?

6. If seven men can do a piece of work in 4 days, how many days would it take one man to do it?

7. If 7 boys earn $7 in 1 day, 1 boy earns $\frac{1}{7}$ of $7, or $ ——.

8. If 7 rows of trees have 77 trees, 1 row has $\frac{1}{7}$ of 77, or —— trees.

9. John buys a cake for 20¢, cuts it into 7 equal pieces, and sells each piece for 5¢. What is his profit?

Liquid Measure — Oral Review

185. 1. By measuring, find how many pints in a quart.

2. Three quarts equal how many pints? 4 quarts? 5 quarts?

3. One pint is what part of a quart?

4. Measure and find how many quarts in a gallon.

5. How many quarts in 3 gallons? 5 gallons? 7 gallons? 9 gallons?

6. What part of a gallon is 1 quart?

7. What part of a gallon are 2 quarts?

8. How many quart cans can you fill from a two-gallon can of sirup?

9. What part of a gallon is a quart?

10. How many pint bottles can be filled with 3 quarts of vinegar?

11. How many times must you fill a pint bottle to measure 2 quarts of milk?

Dry Measure — Oral

186. 1. How many peck measures can be filled from 1 bushel of wheat?

2. A peck is what part of a bushel? What part of a bushel is 2 pecks?

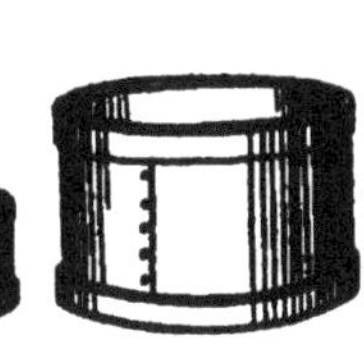

3. If a boy buys a quart of beans, what measure is used?

4. If a girl buys a quart of milk, what measure is used?

5. Peas cost 10¢ a quart. How much will a peck of peas cost?

6. A farmer has 6 pk. of onions. He sold 1 bu. to a grocer. How many pecks has he left?

7. A grocer charges 20¢ a peck for potatoes. What does he charge for a bushel at the same rate?

8. Find the number of quarts in a bushel?

9. How many quarts in a bushel?

10. Estimate the contents of crates, boxes, baskets, and bags to hold garden produce.

DRY MEASURE

8 quarts (qt.) = 1 peck (pk.)
4 pecks (pk.) = 1 bushel (bu.)

Review Exercises — Written

187. 1. Mrs. Jones receives a bill for the following goods:

6 qt. of peas at 10¢ a quart.
7 lb. of coffee at 40¢ a pound.

Find the amount of the bill.

2. At another time Mrs. Jones receives a bill for

15 lb. of sugar at 6¢ a pound.
2 doz. eggs at 43¢ a dozen.

How much must she pay?

3. How many pint bottles can be filled with 19 quarts of milk?

4. How many quart bottles can be filled with 16 gallons of milk?

5. How many pints in 1 gallon?

6. How many pints in 9 gallons?

7. How many pints in 34 gallons?

8. How many quarts of milk does it take to fill 18 pint bottles?

9. How many gallons of sirup does it take to fill 28 quart cans?

188. Find:

1. $\frac{1}{4}$ of 248	2. $\frac{1}{4}$ of 360	3. $\frac{1}{4}$ of 480
4. $\frac{1}{5}$ of 255	5. $\frac{1}{5}$ of 500	6. $\frac{1}{5}$ of 550
7. $\frac{1}{6}$ of 846	8. $\frac{1}{6}$ of 120	9. $\frac{1}{6}$ of 660
10. $\frac{1}{7}$ of 210	11. $\frac{1}{7}$ of 357	12. $\frac{1}{7}$ of 497

Written Problems

189. 1. A roll of wrapping paper is 300 ft. long. How many yards in the roll?

2. A firm employs 7 men. Each man receives a salary of $ 24 a week. How much does the firm pay in salaries in 2 wk.?

3. A school had 343 children on register the first of September. There were 134 children admitted during September and 94 in October. How many children were on register November 1?

4. A man bought a horse for $ 375 and sold it for $ 500. How much did he gain?

5. A man bought a team of horses for $420 and a harness for $\frac{1}{7}$ as much. What did the harness cost?

6. What was the cost of the team and the harness?

7. A handball court is 50 ft. long and 25 ft. wide. How many feet will a boy walk if he walks around the entire court?

8 Make a drawing of the handball court. What part of the length is the width?

9. A farmer has a farm of 160 acres. He sows 60 acres with wheat What part of the farm is the wheat field?

10. A man received a bill containing the following charges: $ 5.58, $ 5.40, $ .70. What is the total amount of the bill?

11. A train is made up of 7 cars. There are 60 seats in each car. Each seat accommodates two persons. How many persons will the train seat?

12. A plot of ground 350 ft. wide is divided into 7 equal parts. How wide is each part?

Multiplication by Eight — Oral

190. Give the answer:

1. Add two 8's. $2 \times 8 = ?$
2. Add three 8's. $3 \times 8 = ?$
3. Add six 8's. $6 \times 8 = ?$
4. Add seven 8's. $7 \times 8 = ?$
5. Add eight 8's. $8 \times 8 = ?$
6. A railroad car is supported by 2 trucks, each with 4 wheels. How many wheels on 8 trucks?

191. Memorize the table:

THE MULTIPLICATION TABLE OF 8	
$1 \times 8 = 8$	$6 \times 8 = 48$
$2 \times 8 = 16$	$7 \times 8 = 56$
$3 \times 8 = 24$	$8 \times 8 = 64$
$4 \times 8 = 32$	$9 \times 8 = 72$
$5 \times 8 = 40$	$10 \times 8 = 80$

Make the table of eight, beginning with 8×1, 8×2, etc.

Division by Eight — Oral

192. 1. How many 8's in 24? 24 ÷ 8 = ?

2. How many 8's in 36? 36 ÷ 8 = ?

3. How many 8's in 16? 16 ÷ 8 = ?

4. How many 8's in 40? 40 ÷ 8 = ?

5. How many 8's in 56? 56 ÷ 8 = ?

6. What is $\frac{1}{8}$ of 16? Of 24?

7. What is $\frac{1}{8}$ of 32? Of 40?

8. $8\overline{)8 \quad 16 \quad 24 \quad 32 \quad 40 \quad 48}$

To find **one eighth** of a number divide the number by **eight**.

193. Study the table:

THE DIVISION TABLE OF 8

8 ÷ 8 = 1	$\begin{array}{r}1\\8\overline{)8}\end{array}$	48 ÷ 8 = 6	$\begin{array}{r}6\\8\overline{)48}\end{array}$
16 ÷ 8 = 2	$\begin{array}{r}2\\8\overline{)16}\end{array}$	56 ÷ 8 = 7	$\begin{array}{r}7\\8\overline{)56}\end{array}$
24 ÷ 8 = 3	$\begin{array}{r}3\\8\overline{)24}\end{array}$	64 ÷ 8 = 8	$\begin{array}{r}8\\8\overline{)64}\end{array}$
32 ÷ 8 = 4	$\begin{array}{r}4\\8\overline{)32}\end{array}$	72 ÷ 8 = 9	$\begin{array}{r}9\\8\overline{)72}\end{array}$
40 ÷ 8 = 5	$\begin{array}{r}5\\8\overline{)40}\end{array}$	80 ÷ 8 = 10	$\begin{array}{r}10\\8\overline{)80}\end{array}$

Drill Device — The Rectangle

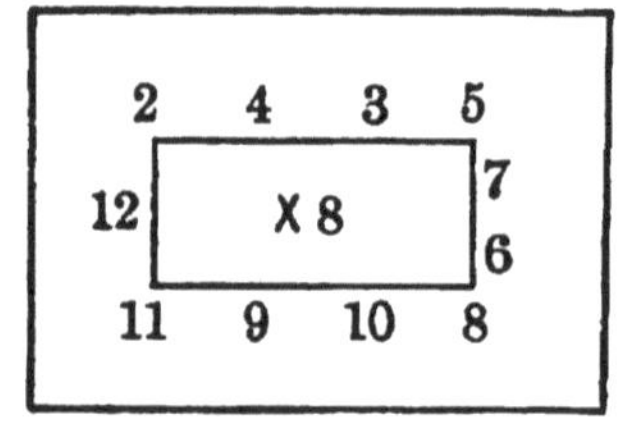

194. Multiply the numbers on the outside by the number in the center.

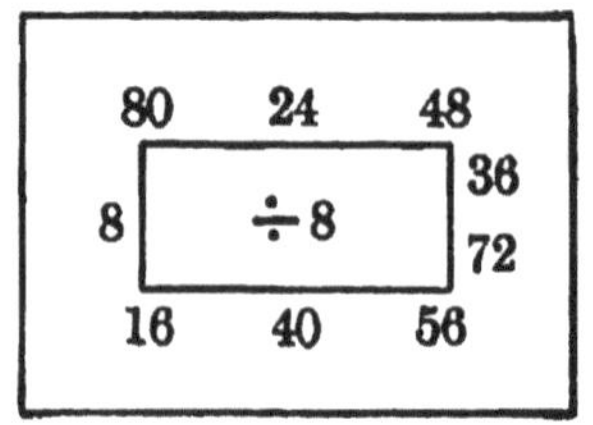

Divide the numbers on the outside by the number in the center.

Multiplication — Written Exercises

195. Find the products:

1.	2.	3.	4.	5.	6.
12 × 8	13 × 8	15 × 8	17 × 8	16 × 8	18 × 8

7.	8.	9.	10.	11.	12.
23 × 8	26 × 8	27 × 8	29 × 8	28 × 8	36 × 8

13. 8 × 137 14. 192 × 8 15. 99 × 8 16. 75 × 8

17. 101 208 309 406 506 602 703 × 8

18. 709 809 121 135 29 209 307 × 8

Division — Written Exercises

196. Find the answers:

1. $\frac{1}{8}$ of 48. 2. $\frac{1}{8}$ of 32. 3. $\frac{1}{8}$ of 40.
4. $\frac{1}{8}$ of 56. 5. $\frac{1}{8}$ of 72. 6. $\frac{1}{8}$ of 64.

7. $8\overline{)88}$ 8. $8\overline{)160}$ 9. $8\overline{)168}$ 10. $8\overline{)248}$

11. $8\overline{)320}$ 12. $8\overline{)408}$ 13. $8\overline{)488}$ 14. $8\overline{)740}$

15. $8\overline{)240}$ 16. $8\overline{)400}$ 17. $8\overline{)888}$ 18. $8\overline{)808}$

Oral Problems

197. 1. If there are 8 panes of glass in a window, how many panes of glass are there in 5 windows? In 8 windows?

2. A railroad car has 8 wheels. How many wheels in a train of 6 cars?

3. How many melons, at 8¢ each, can be purchased for 48¢? For 64¢?

4. How many pounds of sugar, at 8¢ a pound, can be bought for 56¢? For 72¢?

5. A farmhouse of three stories is 24 ft. high. How high is each story?

Written Problems

198. 1. In a gallon there are 8 pints. How many pints in 23 gallons?

2. How many gallons are there in 160 pints?

3. How many pecks are there in 37 bushels?

4. A box has 48 eggs, of which $\frac{1}{4}$ are cracked. How many eggs are cracked? How many are whole?

5. There are 105 lemons in a box. $\frac{1}{5}$ of them are bad. How many lemons are bad? How many are good?

6. At 18¢ a pound, what is the cost of $\frac{1}{2}$ lb. of meat?

7. Measure 8 bushels with a peck measure. How many times must you fill the peck measure?

8. A bushel of beans weighs 60 lb. What will 8 bu. weigh?

9. An employer distributed $168 among his workmen, giving $8 to each. How many workmen did he have?

10. Find the value of each of the following:

8 × $2.05   8 × 35 bu.   8 × 63 lb.   8 × 98 gal.

11. Ned is 11 years old. His grandfather is 8 times as old. How old is his grandfather?

12. A girl paid $1.60 (160¢) for 8 yards of cloth. What was the cost per yard?

13. How many pecks in 2 bushels? In 18 bushels?

14. How many bushels in 48 pecks?

15. What is the cost of 6 copies of a book that sells at $1.80 (180¢)?

Multiplication by 9 — Oral

199. Give the answers:

1. Add two 9's. $2 \times 9 = ?$
2. Add three 9's. $3 \times 9 = ?$
3. Add four 9's. $4 \times 9 = ?$
4. Add five 9's. $5 \times 9 = ?$
5. Add six 9's. $6 \times 9 = ?$
6. Add seven 9's. $7 \times 9 = ?$
7. Add eight 9's. $8 \times 9 = ?$
8. Add nine 9's. $9 \times 9 = ?$

200. Memorize the table:

THE MULTIPLICATION TABLE OF 9	
$1 \times 9 = 9$	$6 \times 9 = 54$
$2 \times 9 = 18$	$7 \times 9 = 63$
$3 \times 9 = 27$	$8 \times 9 = 72$
$4 \times 9 = 36$	$9 \times 9 = 81$
$5 \times 9 = 45$	$10 \times 9 = 90$

Make the table of nine, beginning with 9×1, 9×2, etc.

Division by Nine — Oral

201. 1. Count by 9's to 18. How many 9's in 18? $18 \div 9 = ?$

2. Count by 9's to 27. How many 9's in 27? $27 \div 9 = ?$

3. Count by 9's to 36. How many 9's in 36? $36 \div 9 = ?$

4. Count by 9's to 45. How many 9's in 45? $45 \div 9 = ?$

5. How many 9's in 54? $54 \div 9 = ?$

6. How many 9's in 63? $63 \div 9 = ?$

7. How many 9's in 72? $72 \div 9 = ?$

8. How many 9's in 81? $81 \div 9 = ?$

9. How many 9's in 90? $90 \div 9 = ?$

To find **one ninth** of a number, divide the number by **nine**.

202. Study the table:

THE DIVISION TABLE OF 9

$9 \div 9 = 1$	$\begin{array}{r}1\\ 9\overline{)9}\end{array}$	$54 \div 9 = 6$	$\begin{array}{r}6\\ 9\overline{)54}\end{array}$
$18 \div 9 = 2$	$\begin{array}{r}2\\ 9\overline{)18}\end{array}$	$63 \div 9 = 7$	$\begin{array}{r}7\\ 9\overline{)63}\end{array}$
$27 \div 9 = 3$	$\begin{array}{r}3\\ 9\overline{)27}\end{array}$	$72 \div 9 = 8$	$\begin{array}{r}8\\ 9\overline{)72}\end{array}$
$36 \div 9 = 4$	$\begin{array}{r}4\\ 9\overline{)36}\end{array}$	$81 \div 9 = 9$	$\begin{array}{r}9\\ 9\overline{)81}\end{array}$
$45 \div 9 = 5$	$\begin{array}{r}5\\ 9\overline{)45}\end{array}$	$90 \div 9 = 10$	$\begin{array}{r}10\\ 9\overline{)90}\end{array}$

Drill Device — The Rectangle

203. Multiply the numbers on the outside by the number in the center. Speed.

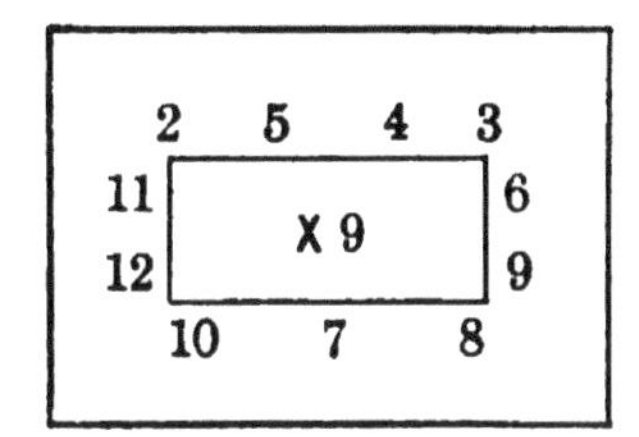

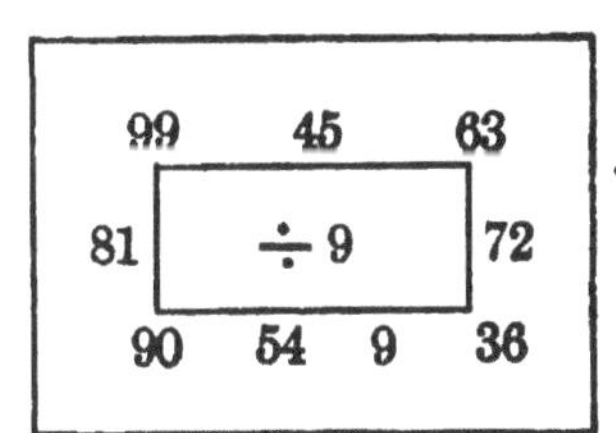

Divide the numbers on the outside by the number in the center. Speed.

Multiplication — Written Exercises

204. Find the product:

1. 11 × 9	2. 12 × 9	3. 13 × 9	4. 14 × 9	5. 15 × 9	6. 16 × 9
7. 17 × 9	8. 18 × 9	9. 19 × 9	10. 23 × 9	11. 25 × 9	12. 27 × 9
13. 28 × 9	14. 29 × 9	15. 45 × 9	16. 72 × 9	17. 99 × 9	18. 75 × 9
19. 101 × 9	20. 203 × 9	21. $1.10 × 9	22. $1.08 × 9	23. $1.23 × 9	
24. 63 bu. × 9	25. 45 yd. × 9	26. 27 mi. × 9	27. 12 in. × 9	28. 14 ft. × 9	

Division — Written Exercises

205. Find the answer:

1. $\frac{1}{9}$ of 18	2. $\frac{1}{9}$ of 27	3. $\frac{1}{9}$ of 36
4. $\frac{1}{9}$ of 45	5. $\frac{1}{9}$ of 54	6. $\frac{1}{9}$ of 63
7. $\frac{1}{9}$ of 81	8. $\frac{1}{9}$ of 72	9. $\frac{1}{9}$ of 81

10. $9\overline{)99}$	11. $9\overline{)180}$	12. $9\overline{)189}$	13. $9\overline{)270}$
14. $9\overline{)279}$	15. $9\overline{)369}$	16. $9\overline{)459}$	17. $9\overline{)549}$
18. $9\overline{)819}$	19. $9\overline{)630}$	20. $9\overline{)639}$	21. $9\overline{)900}$
22. $720 \div 9$	23. $729 \div 9$	24. $999 \div 9$	25. $990 \div 9$

Miscellaneous Problems

206. 1. If I divide 90 violets among 9 girls, how many violets will each receive?

2. How many hours does a man work in a week, if he works 9 hr. a day?

3. If peaches sell 9 for 15¢, what will be the cost of 18 peaches?

4. Cecil is 36 yr. old and John is $\frac{1}{4}$ as old. How old is John?

5. A hotel uses 425 lb. of rice in a week. How many pounds does it use in 9 wk.?

6. A real estate dealer sold a small farm of 9 acres for $ 549. How much did he receive per acre?

7. A farmer has 9 cows. The average daily quantity of milk from each cow is 21 qt. How many

8. A farmer feeds 720 qt. of grain to 9 sheep in one month. How many quarts does he allow for each sheep per month?

9. How many pecks for each sheep?

10. How many bushels for each sheep?

11. A train travels 54 miles per hour. How far does it go in 9 hours?

12. A train runs 369 miles in 9 hours. How far does it run in 1 hour?

13. If you read 27 pages each day, you read —— pages in 9 days.

14. May reads 9 books every month. How many books does she read in a year?

15. At $9 a barrel, how many barrels of flour can be bought for $99?

16. John has 279 stamps, Mary has $\frac{1}{9}$ as many. How many has she?

17. In 9 months, Fred collected 720 stamps. What was the average number collected each month?

18. James has 50¢. He buys a toy hammer and a toy screw driver costing 9¢ each. How much money has he left?

19. What is the cost of 9 yd. of silk at $2.05 a yard?

20. Mr. Hunt has 189 currant bushes in his garden, arranged in 9 rows. How many bushes has he in each row?

21. Teddy has 9 rows of soldiers, with 37 soldiers in each row. How many soldiers has he altogether?

Multiplication by Ten — Oral

207. Give the answers:

1. Add two 10's. $2 \times 10 = ?$
2. Add three 10's. $3 \times 10 = ?$
3. Add four 10's. $4 \times 10 = ?$
4. Add five 10's. $5 \times 10 = ?$
5. Add six 10's. $6 \times 10 = ?$
6. Add seven 10's. $7 \times 10 = ?$
7. Add eight 10's. $8 \times 10 = ?$
8. Add nine 10's. $9 \times 10 = ?$
9. Add ten 10's. $10 \times 10 = ?$
10. What is the last digit in each product?
11. Give an easy way to find the product.

208. Memorize the table:

THE MULTIPLICATION TABLE OF 10	
$1 \times 10 = 10$	$6 \times 10 = 60$
$2 \times 10 = 20$	$7 \times 10 = 70$
$3 \times 10 = 30$	$8 \times 10 = 80$
$4 \times 10 = 40$	$9 \times 10 = 90$
$5 \times 10 = 50$	$10 \times 10 = 100$

Make the table of ten, beginning 10×1, 10×2. etc.

Division by Ten — Oral

209. 1. Count by 10's to 20. $20 \div 10 = ?$

2. Count by 10's to 30. $30 \div 10 = ?$

3. Count by 10's to 40. $40 \div 10 = ?$

4. How many 10's in 50. $50 \div 10 = ?$

5. How many 10's in 60? $60 \div 10 = ?$

6. $10\overline{)80} = ?$ $10\overline{)90} = ?$

7. $10\overline{)100} = ?$ $10\overline{)70} = ?$

8. Give an easy way to find the quotient.

210. Study the table:

THE DIVISION TABLE OF 10

$10 \div 10 = 1$	$\begin{array}{r}1\\10\overline{)10}\end{array}$	$60 \div 10 = 6$	$\begin{array}{r}6\\10\overline{)60}\end{array}$
$20 \div 10 = 2$	$\begin{array}{r}2\\10\overline{)20}\end{array}$	$70 \div 10 = 7$	$\begin{array}{r}7\\10\overline{)70}\end{array}$
$30 \div 10 = 3$	$\begin{array}{r}3\\10\overline{)30}\end{array}$	$80 \div 10 = 8$	$\begin{array}{r}8\\10\overline{)80}\end{array}$
$40 \div 10 = 4$	$\begin{array}{r}4\\10\overline{)40}\end{array}$	$90 \div 10 = 9$	$\begin{array}{r}9\\10\overline{)90}\end{array}$
$50 \div 10 = 5$	$\begin{array}{r}5\\10\overline{)50}\end{array}$	$100 \div 10 = 10$	$\begin{array}{r}10\\10\overline{)100}\end{array}$

To find **one tenth** of a number, divide the number by **ten**.

Drill Device — The Rectangle

211. Multiply the numbers on the outside by the number in the center.

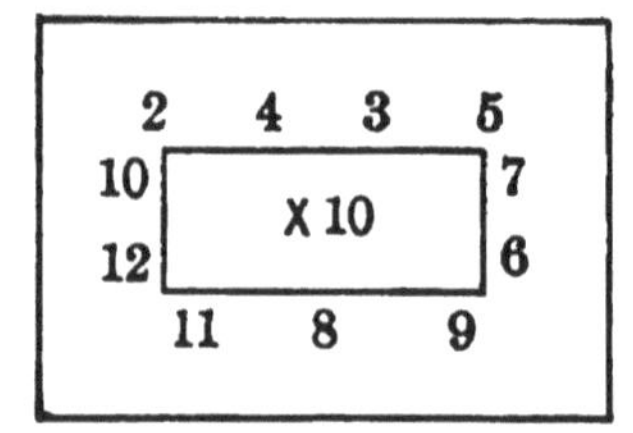

100 80 90
20 ÷10 70
10 60
30 50 40

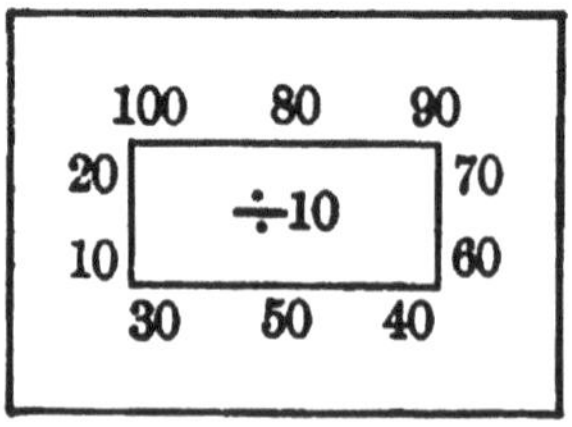

Divide the numbers on the outside by the number in the center.

Oral Exercises — Money

212. 1. How many cents in 5 dimes?

2. How many cents in 10 dimes?

3. How many dimes in 40¢?

4. How many dimes in 60¢?

5. How many dimes in $1.00?

6. What part of $1.00 is 1 dime?

7. What part of 50¢ is 1 dime?

8. What part of 50¢ is 5¢?

9. How many 5-cent pieces in 25¢?

10. What part of 25¢ is 5¢?

Money — Oral Problems

213. 1. A boy bought 50 two-cent stamps. If he paid for them in dimes how many coins did he give?

2. A package of grass seed is sold for 10¢. If 9 packages cost the dealer 50¢, what is his profit on the 9 packages?

3. A pound of candy costs 50¢. How many dimes will pay for 2 lb.?

4. Harry has in his bank 1 half dollar, 2 quarters, 5 dimes, and 2 nickels. How much money has Harry in the bank?

5. James' bank has in it 1 dollar, 3 dimes, 3 nickels. How much money has James in the bank?

6. George bought 10 quarts of berries at 6¢ a quart. He sold them for 9¢ a quart. How much did he gain on one quart? On 9 qt.?

7. 20 dimes are equal to how many dollars?

8. Henry blacked boots in a hotel at 10¢ a pair. How much did he receive from 13 customers?

9. How many cents in 1 dime? How many cents in $1?

10. How many dimes in $1?

Oral Exercises

214. 1. Answer rapidly: 2×10, 5×10, 6×10.

2. What is the last figure in each product?

3. What figure must be written after 7, to get the product 7×10?

To multiply by 10, annex a 0 to the whole number.

4. Answer rapidly $10\overline{)80}$, $10\overline{)70}$.

5. Explain an easy method of dividing a number that ends in 0 by 10.

To divide by 10 a number ending in 0, cancel the 0 at the end.

Review Exercises

215. Tell the products at once:

1. 2×3	2. 5×4	3. 4×6
4. 7×3	5. 9×2	6. 3×9
7. 8×4	8. 6×8	9. 7×6
10. 8×9	11. 6×7	12. 8×7
13. 9×6	14. 8×8	15. 7×7
16. 6×9	17. 9×8	18. 7×10
19. 8×10	20. 9×10	21. 10×10

216. Tell the multiplier and multiplicand that will give each of these products.

1. 24	2 25	3. 27	4. 42
5. 36	6. 28	7. 35	8. 30
9. 49	10. 32	11. 81	12. 96
13. 64	14. 56	15. 54	16. 72
17. 84	18. 63	19. 80	20. 81

217. Supply the missing numbers:

1. $* \times 6 = 48$	2. $7 \times * = 35$	3. $* \times 8 = 56$
$* \times 6 = 42$	$7 \times * = 56$	$* \times 8 = 40$
$* \times 6 = 54$	$7 \times * = 42$	$* \times 8 = 64$
$* \times 6 = 60$	$7 \times * = 63$	$* \times 8 = 48$
$* \times 6 = 36$	$7 \times * = 70$	$* \times 8 = 80$

4. $9 \times * = 63$
$9 \times * = 72$
$9 \times * = 45$
$9 \times * = 54$
$9 \times * = 81$

5. $* \times 10 = 50$
$* \times 10 = 70$
$* \times 10 = 100$
$* \times 10 = 80$
$* \times 10 = 90$

6. $* \times 6 = 54$
$7 \times * = 56$
$* \times 6 = 72$
$9 \times * = 63$
$10 \times * = 60$

A Table for Reference in Multiplication

218. To find the product of any two numbers by this table, as 5×4, find 5 in the first column. Then find the number to the right of 5 in the column of 4.

The square gives the multiplication table as far as we have gone. Rule 100 little squares and write the table from memory.

1	2	3	4	5	6	7	8	9	10
2	4	6	8	10	12	14	16	18	20
3	6	9	12	15	18	21	24	27	30
4	8	12	16	20	24	28	32	36	40
5	10	15	20	25	30	35	40	45	50
6	12	18	24	30	36	42	48	54	60
7	14	21	28	35	42	49	56	63	70
8	16	24	32	40	48	56	64	72	80
9	18	27	36	45	54	63	72	81	90
10	20	30	40	50	60	70	80	90	100

1. Show from this table that 2×5 is the same as 5×2.

2. Show that $3 \times 7 = 21 = 7 \times 3$, $7 \times 9 = 63 = 9 \times 7$.

3. This table gives $9 \times 10 = 90$. Is this correct? Multiply and see.

Business Application

219. 1. How many nickels are in this rack? How many quarters? How many dimes?

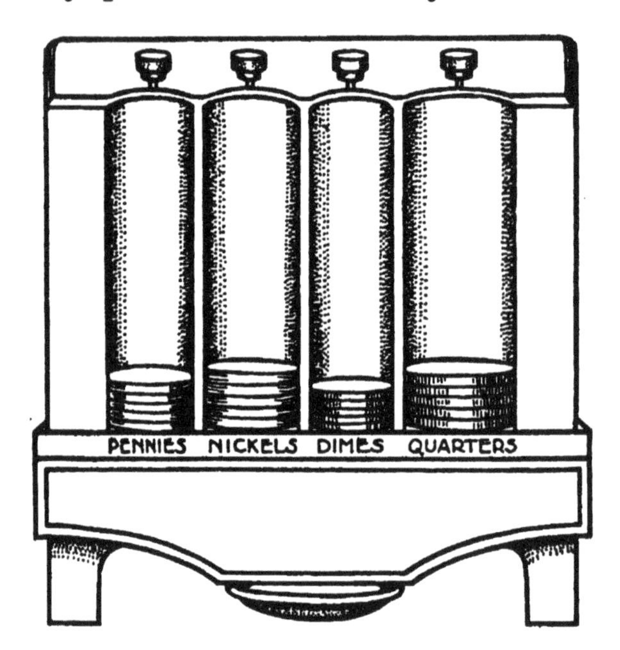

2. The coin rack is a convenient means of making change. Explain its use.

3. How many cents are worth 2 nickels? 4 nickels? 7 nickels? 9 nickels? 10 nickels?

4. 5 dimes are worth how many nickels? How many cents?

5. How many cents in 9 dimes? 8 dimes? 7 dimes?

6. How many nickels are worth 7 quarters, 2 quarters, 3 quarters?

7. If 3 dimes are taken out of the rack, how many nickels must be put in to make up the amount?

8. If 2 quarters are taken out, how many dimes must be put in to make up the amount?

9. The total amount is not changed, if you take out 1 quarter and put in —— dimes and —— nickels.

10. If you take out 6 dimes and 3 nickels, how many quarters must be put in to make up the amount?

11. Make a problem in which the change may be taken from the coin holder.

12. Explain how the coin holder is used in business transactions.

REVIEW

220. 1. $3 \times 10 + 37 = ?$

PROCESS

$$\begin{array}{r} 30 \\ +37 \\ \hline 67 \end{array}$$ *Ans.*

EXPLANATION. — In examples like this *first multiply and then add or subtract.*

$3 \times 10 = 30$, $30 + 37 = 67$, *Ans.*

2. $4 \times 11 - 26 = ?$

PROCESS

$$\begin{array}{r} 44 \\ -26 \\ \hline 18 \end{array}$$ *Ans.*

EXPLANATION

$4 \times 11 = 44$, $44 - 26 = 18$, *Ans.*

3. $2 \times 10 + 43 = ?$
4. $3 \times 10 + 45 = ?$
5. $6 \times 10 + 29 = ?$
6. $4 \times 10 + 65 = ?$
7. $5 \times 10 + 115 = ?$
8. $7 \times 10 + 15 = ?$
9. $8 \times 3 - 5 \times 3 = ?$
10. $3 \times 11 + 6 \times 11 = ?$
11. $8 \times 8 - 7 \times 8 = ?$
12. $7 \times 7 + 3 \times 7 = ?$
13. $7 \times 8 - 7 \times 3 = ?$
14. $6 \times 8 + 4 \times 8 = ?$

Written Exercise

221. Find the answer:

1. 12 21 32 33 44 55 $\times 10$

2. 61 72 83 94 57 46 $\times 10$

3. 16 27 38 49 75 64 $\times 10$

4. $10\overline{)100}$ 120 130 140 150

5. $10\overline{)200}$ 320 440 550 660

Drill Exercises — A Tournament

222. 1. A party of boys held a tournament. A marble shot through an arch of the bridge counted a gain of the number of points marked over the arch. If a shot did not go through an arch, it counted a loss of 5 points. The records of 10 trials for each boy are given below. What was each boy's score? Who won?

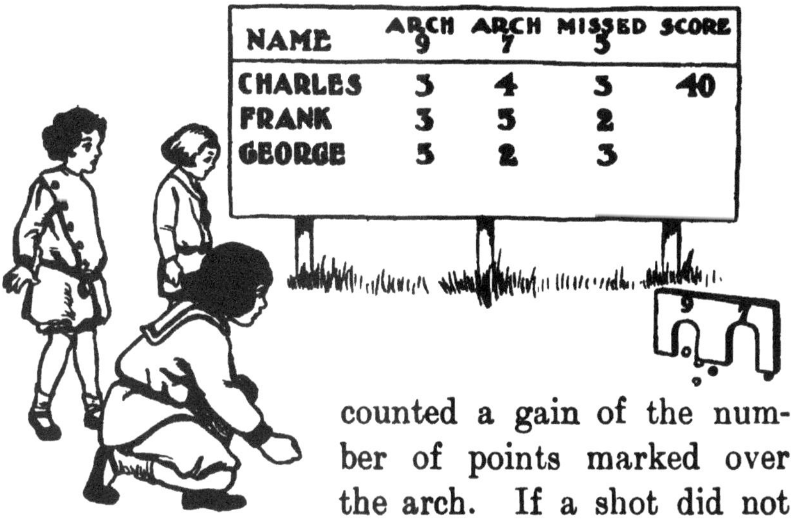

Name	Arch 9	Arch 7	Missed 5	Score
Charles	3	4	3	40
Frank	3	5	2	
George	5	2	3	
Henry	4	2	4	
James	2	4	4	
John	0	9	1	
Ned	4	4	2	
Paul	5	2	3	

PROCESS

$$\begin{array}{rl} & 27 \\ + & \underline{28} \\ & 55 \text{ Gain} \\ - & \underline{15} \text{ Loss} \\ & 40 \text{ Score} \end{array}$$

EXPLANATION. — Charles' score was $3 \times 9 + 4 \times 7 - 3 \times 5$.

2. Put over the arches the figures 6 and 8. Prepare a new record of ten trials. Find the score of each boy.

Continue the game.

Division — Remainders

223. Divide:

1. 7 by 2.

PROCESS

$$\begin{array}{r} 3 \\ 2\overline{)7} \\ \underline{6} \\ 1 \end{array}$$

EXPLANATION. — $7 \div 2 = 3$ and a remainder. Write the 3.

$2 \times 3 = 6$. Write the 6 beneath the 7. Subtract.

There is a remainder 1.

The quotient is 3, and the remainder is 1.

Check the answer.

224. **Written Exercises**

1. $2\overline{)5}$	2. $2\overline{)9}$	3. $2\overline{)11}$	4. $2\overline{)13}$	5. $2\overline{)15}$
6. $2\overline{)17}$	7. $2\overline{)19}$	8. $3\overline{)5}$	9. $3\overline{)7}$	10. $3\overline{)4}$
11. $3\overline{)8}$	12. $3\overline{)10}$	13. $3\overline{)11}$	14. $3\overline{)13}$	15. $3\overline{)14}$

16. 3)16	17. 3)17	18. 3)20	19. 3)19	20. 3)22
21. 3)23	22. 3)25	23. 3)26	24. 3)28	25. 3)29
26. 4)7	27. 4)11	28. 4)17	29. 4)19	30. 4)21
31. 5)7	32. 6)8	33. 7)9	34. 8)10	35. 9)21
36. 8)18	37. 9)28	38. 9)30	39. 10)11	40. 4)35

In review, the foregoing exercises may be worked orally.

Problems

225. 1. Will has 7 pears and gives his 3 sisters 2 pears each. How many pears are left for him?

2. James has 13 apples. He gives 2 apples to each of 5 boys. How many apples are left for him?

3. A milkman has 16 pints of milk. He delivers to 7 customers 2 pints each. How many pints has he then left?

4. Mary had 13¢ in her bank and gave 2¢ every Sunday to Sunday school. How many Sundays did the money last? What remained?

5. How many two-cent postage stamps can a boy purchase with 19¢? How many cents has he left?

6. How many gallon cans can be filled from a can holding 18 quarts?

7. In measuring a 14-ft. wall with a yard-stick, how many feet are left over after the whole number of yards has been measured?

8. How many yards and feet in 17 ft.?

9. At 6¢ apiece, how many notebooks can be purchased with 15¢?

10. A teacher divided her class of 18 pupils into groups of 4. How many whole groups were there?

11. In a school there are four rooms. In the first room there are 25 children, in the second 27, in the third 21, in the fourth 15. If the pupils were distributed so that all the rooms had the same number of children, how many would there be in each room?

12. A farmer raised each year for four years 8 acres of wheat. The total number of bushels raised the first year was 80, the second year 88, the third year 96, and the fourth year 96. What was the total yield for the four years? What was the average yield per acre for each year?

MULTIPLICATION AND DIVISION

Written Exercises

226. Divide:

1. 54 by 2.

PROCESS

```
 t. o.
  27
2)54
  4
  14
  14
```

EXPLANATION. — $5 \div 2 = 2$ and a remainder. Write the 2 above the line over the 5 in tens' place.

Multiply and subtract.

The remainder in tens' place is 1. Bring down the 4 from ones' place.

The new dividend is 14. $14 \div 2 = 7$.

Multiply and subtract.

There is no remainder. The quotient is 27.

2. 2)32	3. 2)52	4. 2)72	5. $90 \div 2$
6. 2)92	7. 3)42	8. 3)72	9. $45 \div 3$
10. 2)92	11. 4)96	12. 3)78	13. $57 \div 3$
14. 4)52	15. 4)60	16. 4)68	17. 5)60
18. 5)65	19. 6)72	20. 6)84	21. 6)90
22. 6)96	23. 7)84	24. 7)91	25. 8)96
26. 8)104	27. 9)108	28. 9)117	29. 7)175

Written Problems

227. 1. At $ 5 each, how many sheep can be bought for $ 65?

2. If 5 pigeons cost 95¢, 1 pigeon cost $\frac{1}{5}$ of 95¢, or —— cents.

3. Jane has 42 yd. of ribbon which she cuts into 3-yd. lengths. How many lengths has she?

4. How many boxes of biscuit, at 7¢ each, can be bought for 98¢?

5. How many bunches of celery having 4 stalks in each bunch can be made from 76 stalks?

6. There are 136 car wheels in a railroad shop. If each car requires 8 wheels, how many cars can be supplied?

7. If $84 is divided equally among 6 boys, how much does each receive?

8. Which is the cheaper — to buy tablets, 6 for 90¢, or 7 for 98¢?

9. A teacher has 60 sticks of chalk. If the class uses 4 sticks each day, how many days will the chalk last?

10. A teacher gives 72 sheets of paper to her class. If she gives 3 sheets to each pupil, how many pupils are there in the class?

Written Exercises

228. Which is the greater:

1. $\frac{1}{5}$ of 75 or $\frac{1}{6}$ of 84?
2. $\frac{1}{7}$ of 84 or $\frac{1}{8}$ of 104?
3. $\frac{1}{6}$ of 48 or $\frac{1}{9}$ of 63?
4. $\frac{1}{4}$ of 88 or $\frac{1}{6}$ of 96?
5. $\frac{1}{7}$ of 84 or $\frac{1}{6}$ of 78?

Review. Written Exercises

229.

1. 4×41	2. 5×62	3. 7×73
4. 7×84	5. 8×95	6. 9×71
7. 83×5	8. 47×68	9. 38×7
10. 401×5	11. $3\overline{)81}$	12. $4\overline{)96}$
13. $4\overline{)88}$	14. $3\overline{)69}$	15. $5\overline{)65}$
16. $6\overline{)72}$	17. $7\overline{)126}$	18. $7\overline{)147}$
19. $8\overline{)488}$	20. $3\overline{)63}$	21. $9\overline{)909}$
22. $9\overline{)810}$	23. $8\overline{)96}$	24. $7\overline{)637}$
25. $\begin{array}{r} 54 \\ +18 \\ \hline \end{array}$	26. $\begin{array}{r} 17 \\ +11 \\ \hline \end{array}$	27. $\begin{array}{r} 28 \\ -17 \\ \hline \end{array}$
28. $\begin{array}{r} 28 \\ -17 \\ \hline \end{array}$	29. $\begin{array}{r} 16 \\ +13 \\ \hline \end{array}$	30. $\begin{array}{r} 29 \\ -16 \\ \hline \end{array}$

31. $10\overline{)170}$ 32. $10\overline{)100}$ 33. $10\overline{)80}$

34. $8\overline{)80}$ 35. $10\overline{)90}$ 36. 11×9

37. $11\overline{)99}$ 38. $9\overline{)99}$ 39. $10\overline{)240}$

40. $10\overline{)40}$ 41. 11×10 42. $10\overline{)110}$

43. $11\overline{)110}$ 44. 24×10 45. 10×24

46. 10×10 47. 18×10 48. $10\overline{)180}$

49. $10\overline{)150}$ 50. $10\overline{)190}$ 51. $10\overline{)730}$

Oral Problems

230. 1. At $ 5.00 a pair, what is the cost of 8 pairs of shoes?

2. What is the cost of one pair of boots if 8 pairs cost $ 40.00?

3. A merchant sells 9 coats at $8.00 each. How many dollars does he receive?

4. If a merchant sells 9 coats for $ 72.00, how many dollars does he charge for each?

5. There are 5 desks in a row. How many desks in 7 rows?

6. If there are 7 panes in a window, how many panes in 6 windows?

7. If 9 windows contain 54 panes, how many panes are there in each window?

8. How much will 7 heads of lettuce cost at 5¢ apiece?

9. How many oranges, at 5¢ apiece, can be bought for 45¢?

10. What will 9 heads of cabbage cost at 9¢ a head?

Reading Time — Oral Exercise

231. 1. Point out the hands on the face of the clock.

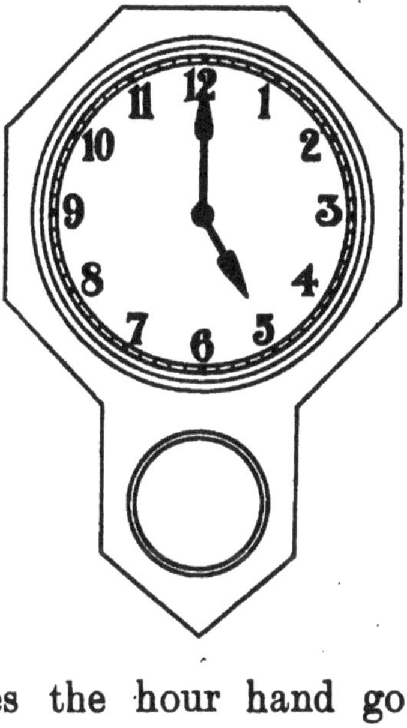

2. Which is the hour hand?

3. Which is the minute hand?

4. Which moves the faster?

5. Show how far the hour hand moves in one hour.

6. Show how far the minute hand moves in one hour.

7. How many times does the hour hand go around in an entire day?

8. How many times does the hour hand go around in 12 hr.?

9. How many times does the minute hand go around in half a day? In a whole day?

10. How many hours are there between breakfast at 7 o'clock and luncheon at 12 o'clock?

11. How many hours between the opening and the closing of your school?

12. How many minutes in half an hour? In one fourth of an hour?

13. If in 1 hour there are 60 minutes, how many minutes are there in $1\frac{1}{2}$ hr.?

14. How many hours in $\frac{1}{2}$ day?

15. How many hours in $1\frac{1}{2}$ days?

16. How many hours in one third of a day?

17. If a man works 8 hr. a day, what part of a day does he work?

18. A boy begins work at 8 in the morning and stops at 1 in the afternoon. How many hours a day does he work? How many hours in 6 da.?

19. Name the months of the year. How many are there?

20. Frank went to school 6 months of one year. What part of the year did he attend?

21. How many months in $\frac{1}{2}$ year? In $\frac{1}{3}$ year? In $\frac{1}{4}$ year?

22. What part of the year are 3 months? 4 months?

23. How many months in $1\frac{1}{2}$ years?

24. How many hours in one fourth of a day?

25. What part of a day are 6 hr.? 12 hr.? 8 hr.?

26. Mary read from 9 o'clock in the morning until 1 in the afternoon. What part of the day did she spend reading?

27. What is the number of hours in 10 days?

28. How many minutes in 1 hour? In 10 hours?

29. How many days in 24 hr.? In 240 hr.?

REVIEW

232. 1. Julia spends 10¢ for candy, 20¢ for a doll, and 30¢ for a book. How much does she spend in all?

2. John read on Friday 10 pages in a history book, on Saturday 30 pages, and on Sunday 50 pages. How many pages did he read in the three days?

3. A hunting party traveled 10 miles the first day, 10 miles the second day, and 20 miles the third day. How many miles in the three days?

4. John earned 70¢ the first day, spent 60¢ the second day, and earned 80¢ the third day. How much did he have at the close of the third day?

5. At 5¢ a ride, how many times can I ride in the street car for 35¢?

6. Charles has 5 cents, Robert has 4 times as many cents as Charles. How many cents have both?

7. Helen can read 3 pages of her book in 18 minutes. In how many minutes can she read 9 pages?

8. Herbert invited 14 playmates to a picnic. $\frac{1}{7}$ of them remained away. How many came?

9. A grocer bought a chest of tea for $ 28.00 and sold it for $ 37.00. What was his gain?

10. A man drove 45 miles in 5 hours. How far did he drive in 1 hour?

11. A boy had 16 squabs. He bought 9 more and then sold 15. How many had he left?

12. The product of two numbers is 42. One of the numbers is 7. What is the other?

13. At 30¢ a yard, what does 1 foot of braid cost?

14. At 10¢ a basket, how many baskets of apples will one dollar buy?

15. How many lead pencils at 5¢ each can I buy for 40¢?

Review — Written Exercises

233.

1.	2.	3.	4.	5.	6.
2	3	6	8	5	7
9	1	4	5	7	5
7	8	2	2	9	3
5	6	9	7	2	1
3	4	6	4	4	9
1	2	3	1	6	7

7.	8.	9.	10.	11.	12.
8	4	5	8	6	4
5	7	4	9	7	5
4	3	3	2	8	6
9	9	2	3	9	4
8	2	6	4	2	5
7	8	7	5	3	8

13.	14.	15.	16.
1	9	31	23
2	8	42	91
3	7	53	34
4	6	64	82
5	5	75	49
6	4	76	56
7	3	87	37

17.	18.	19.	20.
82	73	55	28
19	90	78	45
94	81	69	48
71	34	26	36
12	27	90	15
45	83	12	90
38	25	93	67

21.	22.	23.	24.
642	209	746	392
748	892	577	293
920	740	690	923
148	609	812	329

25.	26.	27.	28.
765	$1.02	$1.96	$2.37
675	.96	1.10	.87
567	1.13	.45	4.32
756	.67	.87	1.23

29.	30.	31.
$20.63	$67.30	$ 7.45
.48	.75	109.55
35.00	4.47	87.60
7.23	36.05	59.35

32. 307 + 2948 + 21 = ?

33. 176 + 23 + 119 + 300 = ?

34. 48 + 57 + 65 + 89 = ?

35.	36.	37.	38.
314	209	312	987
– 187	– 116	– 147	– 789

39.	40.	41.	42.
723	861	429	615
– 246	– 743	– 237	– 309

43.	44.	45.	46.
427	532	620	740
– 392	– 238	– 289	– 375

47.	48.	49.	50.
530	860	780	708
– 162	– 448	– 538	– 246

51.	52.	53.	54.
805	407	603	309
– 532	– 368	– 179	– 271

55.	56.	57.	58.
$4.00	$5.00	$6.00	$7.00
– 2.45	– 3.56	– 4.77	– 5.78

Review — Written Problems

234. 1. A boy's entire collection of stamps contains 1,965 stamps. He has 253 stamps that are duplicates. How many stamps has he that are not duplicated?

2. In a collection of 2,100 stamps there are 172 United States stamps, 74 Canadian stamps, 3 Australian stamps, 12 German stamps, 5 Austrian stamps, and 10 French stamps. How many stamps of other denominations has he?

3. Harry has 973 stamps. If they are worth 1 cent apiece, what is their value in dollars and cents?

4. George bought 213 stamps at 2¢ apiece. How much did he pay?

5. Herbert paid 9¢ apiece for 73 stamps. How much did they cost him?

6. John sells 32 stamps at 6¢ each. How much does he receive? He buys 17 stamps at 8¢ each. How much does he pay for them?

7. Of 56 tomato plants that Ned planted, 37 lived. How many died?

8. I bought 2 horses for $215. For what must I sell them to gain $10 on each?

9. A man bought a dozen brooms at 36¢ each. He sold them at 45¢ each. What was his gain on the dozen?

10. Six loaves of bread cost a baker 36¢. He sold them for 60¢. What is his gain on each loaf of bread?

11. Mr. Knapp pays $ 19.00 a month rent. How much does he pay in 4 months?

12. In Mr. Hull's orchard there are 7 rows of trees, and there are 14 trees in each row. How many trees are there in the orchard?

13. How many eggs are there in 8 dozen?

14. A farmer goes to market five times a week, a distance of 15 miles each way. How many miles does he travel in a week?

15. Mr. Brown, Mr. Hall, and Mr. Roberts cultivated a vegetable garden. Besides supplying their families, they sold vegetables to the amount of $ 63.00, which they divided equally. How much did each receive?

16. How many gallons are there in 176 quarts?

17. The rim of a wheel of a bicycle is 8 ft. around. How far will the bicycle travel in 25 revolutions?

18. How far does a boy travel on a bicycle in 1 hr., if he goes 1 mi. in 6 min.?

19. $\frac{1}{2}$ of a tree 48 ft. high was broken off. How many feet were left standing?

20. If 1 inch on a map stands for 2 miles, how many inches stand for 36 miles?

21. James picked 19 qt. of berries in the forenoon and 23 qt. in the afternoon. How many quarts did he pick?

22. A farmer sold 16 qt. of cherries at 6¢ a quart. How much did he receive?

23. If a man works 8 hr. a day, how many hours does he work in 6 days? In 12 days?

24. A man buys 2 lb. of beefsteak at 22¢ a pound and gives the butcher $1. How much change will he receive?

25. Trees are planted in a row 25 feet apart. A boy starts from a certain tree and passes 9 others. How far has he walked?

PART TWO

READING AND WRITING NUMBERS

Oral Exercise

235. Read these numbers and tell what each figure means:

1. 11	2. 111	3. 1,111	4. 11,111
5. 22	6. 222	7. 2,222	8. 22,222
9. 55	10. 555	11. 5,555	12. 55,555
13. 99	14. 999	15. 9,999	16. 99,999
17. 25	18. 250	19. 2,525	20. 25,251
21. 39	22. 395	23. 3,900	24. 39,000
25. 10	26. 100	27. 1,000	28. 10,000

Oral Review

236. 1. How many units, or ones, make one ten?

2. How many tens make one hundred?
3. How many hundreds make one thousand?
4. How many thousands make ten thousand?
5. What number is one more than 9?
6. What number is one more than 99?
7. What number is one more than 999?

Written Exercise

237. Write these numbers in figures:

1. One thousand.
2. Five thousand.
3. Nine thousand.
4. Ten thousand.
5. Ninety-nine thousand.
6. Nine hundred.

Exercise

238. Point out on the chart the columns in which to write the digits of the following numbers:

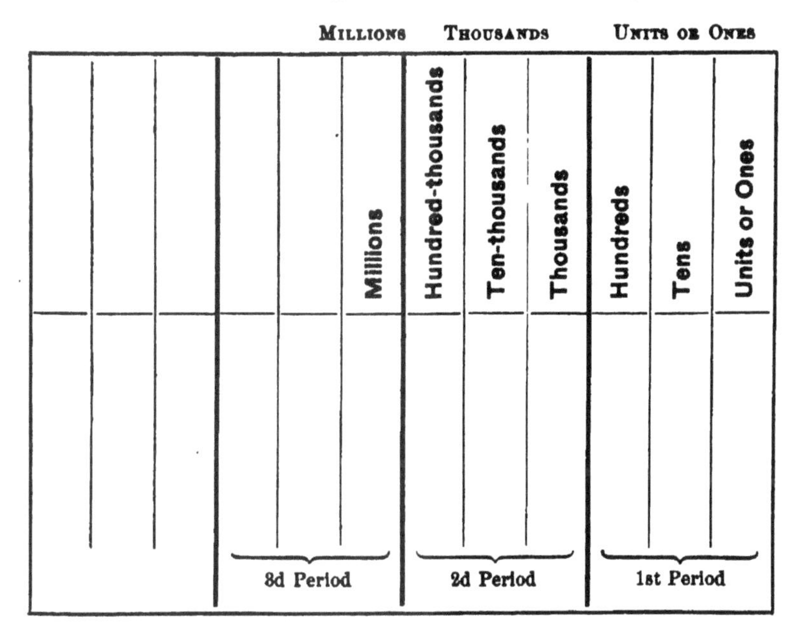

1. 70
2. 700
3. 9,000
4. 49,000
5. 275,000
6. 900,001
7. 290,092
8. 459,023
9. 550,975
10. 654,321
11. 543,216
12. 432,165

Read the numbers.

What is the largest digit that may be written in a column?

Exercise

239. Draw a chart showing the columns in the orders of ones and thousands.

On the chart write the following numbers, filling in the vacant places in the number with ciphers.

1. 4 in hundred-thousands, 5 in hundreds, 3 in ones.
2. 6 in hundred-thousands, 7 in ten-thousands, 4 in tens.
3. 9 in hundred-thousands, 4 in thousands, 3 in hundreds.
4. 7 in ten thousands, 8 in tens, 5 in ones.
5. 5 in ten-thousands, 9 in ones.

Read the numbers. Illustrate as far as possible with bundles of splints.

Exercise

240. Draw a chart similar to the chart in Art. 238. Write the following numbers, filling in vacant places with ciphers. Read the numbers.

1. 9 in tens column.
2. 7 in ten-thousands column.
3. 5 in hundred-thousands column.
4. 3 in tens column.
5. 6 in ten-thousands column.
6. 8 in hundred-thousands column.

Oral Exercise

241. Read the numbers:

1. 4,321	2. 10,000	3. 20,060
4. 54,321	5. 406,050	6. 431,065
7. 654,321	8. 46,050	9. 631,105
10. 604,301	11. 320,000	12. 105,690

In reading large numbers it is convenient to separate the digits of whole numbers, from right to left, into groups of three figures. These groups are called periods. Point out the periods in the numbers in this exercise.

Written Exercise

242. Write in figures:

1. Five thousand sixty.
2. Eight thousand five hundred.
3. Six hundred ninety.
4. Two thousand one.
5. One thousand fifty-five.
6. Five hundred thousand five hundred.
7. Six thousand two hundred eighty-four.
8. Ninety-six thousand eight hundred forty-seven.
9. Forty thousand one hundred seventy-four.
10. Three hundred thirty-one thousand.
11. Five hundred thousand three hundred.
12. One hundred seventy-six thousand.

ADDITION AND SUBTRACTION

243. Add:

1.	2.	3.	4.	5.	6.
65	62	65	25	48	101
43	23	71	66	77	56
46	45	100	80	46	45
81	21	15	33	89	37
50	93	18	17	23	88
104	97	59	84	13	15
53	40	34	19	70	92

7.	8.	9.	10.
7,261	5,412	41,228	72,839
407	8,763	9,347	64,984
4,928	1,498	82,556	86,092
5,654	7,243	29,462	38,677
525	5,398	8,014	97,006
708	7,576	57,198	40,009
6,241	1,417	6,249	35,787
937	302	84,968	65,090

Written Exercise

244. Add by lines and by columns. Check.

1.	2.
18 + 30 + 55 + 44 =	147 + 342 + 291 =
37 + 41 + 8 + 9 =	634 + 459 + 728 =
54 + 92 + 13 + 24 =	963 + 897 + 99 =
33 + 84 + 7 + 17 =	489 + 382 + 965 =
72 + 73 + 96 + 80 =	377 + 648 + 49 =

3.

$7.97 +	$21.01 +	$34.10 +	$294.63 =	
9.15 +	17.63 +	14.05 +	897.32 =	
8.19 +	6 74 +	15.00 +	29.75 =	
6.83 +	40.44 +	2.40 +	.40 =	
4.24 +	9.67 +	42.85 +	600.00 =	
+	+	+	=	

4.

100 +	98 +	213 +	634 +	21 =
7 +	18 +	27 +	459 +	48 =
1,001 +	2,113 +	1,202 +	12 +	395 =
+	+	+		=

5.

404 +	708 +	604 +	909 +	879 +	594 =
35 +	43 +	109 +	749 +	968 +	789 =
124 +	3,120 +	4,321 +	9876 +	428 +	39 =
+	+	+	+	+	=

Subtraction — Written Exercise

245. Subtract:

1.	2.	3.	4.	5.
4,032	6,074	8,250	7,601	6,075
2,873	2,168	5,332	2,172	4,386

6.	7.	8.	9.	10.
4,077	9,068	7,027	4,660	8,560
2,699	3,589	4,279	2,818	3,789

11.	12.	13.	14.	15.
5,000	6,000	7,000	8,000	9,000
4,175	5,287	6,349	7,924	8,876

Subtraction by the Austrian Method.

PROCESS	EXPLANATION
4,032	Then $3 + 9 = 12$, write the 9
2,873	$7 + 5 = 12$, write the 5
1,159	$8 + 1 = 9$, write the 1
	$2 + 1 = 3$, write the 1

Measures of Length — Exercise

246. Measure with the yardstick:

1. The length of the room. Write the whole number in the answer in yards and the remainder in feet and inches.

2. The width of the room. Write the answer in the same way.

3. The length of the table.

4. The length of the blackboard.

5. The length of the school yard.

6. How many feet in 5 yd.? In 8 yd.?

7. How many yards in 30 ft.? In 24 ft.?

8. How many feet in half a yard?

9. How many inches in half a yard?

10. How many inches in half a foot?

11. How many inches in a fourth of a foot?

12. How many feet in a third of a yard?

13. How many inches in a third of a yard?

14. How many inches in a third of a foot?

247.

MEASURE OF LENGTH	
12 inches (in.)	= 1 foot (ft.)
3 feet	= 1 yard (yd.)
5,280 feet	= 1 mile (mi.)

Measure with the rule distances as follows:

1. 3 yd.
2. 4 yd.
3. 1 yd. 2 ft.
4. 1 yd. $1\frac{1}{2}$ ft.
5. 1 yd. 1 ft. 6 in.
6. $\frac{1}{2}$ yd.
7. 2 yd. 2 ft. 2 in.
8. 4 yd. 6 in.

Estimating Distances

248. 1. Estimate the length of the cover of your arithmetic in inches.

2. The length of your desk in yards.

3. The length of the building in yards.

4. The distance between your home and the school in miles.

5. The distance from your house to the post office in miles.

6. Measure as nearly as possible the distances estimated.

Measure of Time — Exercise

249. 1. How many minutes in an hour?

2. How many hours in a day?

3. How many days in a week?

4. How many months in a year?
5. How many seconds in a minute?

TABLE OF TIME

60 seconds (sec.)	= 1 minute (min.)
60 minutes	= 1 hour (hr.)
24 hours	= 1 day (da.)
7 days	= 1 week (wk.)
4 weeks	= 1 month (mo.)
12 months	= 1 year (yr.)

Exercise

250. 1. How many minutes in half an hour?

2. How many minutes in one fourth of an hour?
3. How many hours in half a day?
4. How many hours in a third of a day?
5. How many days in a week?
6. How many days in half a week?
7. How many weeks in half a month?
8. How many weeks in a quarter of a month?
9. How many months in half a year?
10. How many months in a quarter of a year?

Oral Problems

251. 1. It takes 20 min. to walk a mile. How long does it take to walk 2 mi.? 3 mi.?

2. If Richard rides on a bicycle 12 mi. in 1 hr., how far does he go in 4 hr.?

3. At 12 mi. an hour, how long does it take to go 36 mi.?

4. James rides 11 mi. in 1 hr., or —— mi. in 6 hr.

5. At 11 mi. an hour, how long does it take to go 33 mi.?

6. At 10 mi. an hour, how long does it take to go 50 mi.?

7. A man travels 36 miles in 12 hours. How many miles does he travel in 1 hour?

8. At the rate of 9 miles an hour, how far can one travel in 11 hours? In 12 hours?

9. A man travels a distance of 35 mi., going 5 mi. an hour. How long does it take him to travel this distance?

10. A driver runs an automobile 30 mi. in 2 hr. How many miles does he travel in 1 hr.?

Multiplication and Division by Eleven — Oral Exercise

252. Give the products:

1. 2×11	2. 3×11	3. 4×11
4. 5×11	5. 6×11	6. 7×11
7. 8×11	8. 9×11	9. 10×11

Make and memorize the multiplication table of eleven.

253. Count:

1. By 11's to 66, beginning with 11.
2. By 11's to 110, beginning with 11.
3. By 11 from 99 back to 11.
4. By 11's to 78, beginning with 1.
5. To 103, by 11's, beginning with 4.

254. Give the quotients:

1. $11\overline{)22}$ 2. $11\overline{)33}$ 3. $11\overline{)55}$ 4. $11\overline{)121}$ 5. $11\overline{)77}$

6. $11\overline{)99}$ 7. $11\overline{)44}$ 8. $11\overline{)88}$ 9. $11\overline{)110}$ 10. $11\overline{)132}$

Oral Problems

255. 1. If 1 gal. of oil costs 11¢, what will 9 gal. cost? 7 gal.? 10 gal.? 6 gal.?

2. How many quarts in 11 gal.? How many pints?

3. If 11 yd. of tape cost 22¢, what is the cost of 1 yd.?

4. If 11 pencils cost 33¢, 1 pencil will cost ——¢.

5. In a schoolroom there are 44 seats. If there are 11 seats in a row, how many rows are there?

6. An office building is 5 stories high. There are 11 windows in each story. How many windows in the building?

7. There are 132 trees in an orchard set in 11 rows. How many trees in each row?

8. What will 8 yd. of cloth cost at 11 ¢ a yard?

9. A school has 5 football teams. There are 11 players on each team. How many players on the 5 teams?

10. A storekeeper paid $ 110 for 11 coats. What is the average cost of a coat?

Drill Device — The Circle

256. Multiply the numbers on the circle by the number in the center.

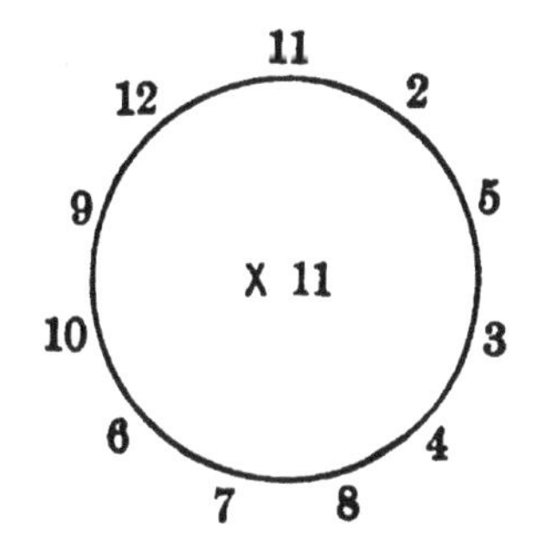

Make a drill device for division by 11. Use on the device the numbers 44, 33, 99, 77, 88, 55, 66, 22. Drill if necessary. What numbers less than 100 are exactly divisible by 11?

Multiplication and Division

257. Multiply:

1. 469 by 11.

PROCESS

```
  469
   11
  ---
  469
 4 69
 ----
5,159 Ans.
```

EXPLANATION. — Multiply first by the 1 in ones' place.

$1 \times 469 = 469$. Write this down.

469×1 *ten* $= 469$ *tens*

Write down 469, placing the 9 in *tens'* place, under the 6.

Adding the two partial products gives 5,159.

This sum is the answer.

Divide:

2. 198 by 11.

PROCESS

```
    18 Ans.
11)198
   11
   --
    88
    88
    --
     0
```

EXPLANATION.—$19 \div 11 = 1$ and a remainder.

Write the 1 in *tens'* place, above the 9.

$1 \times 11 = 11$. Write the 11 underneath the 19.

Subtracting, the remainder is 8 *tens*.

8 *tens* + 8 *ones* = 88 *ones*.
88 *ones* ÷ 11 = 8 *ones*.
Write the 8 in *ones'* place, above the dividend.
$8 \times 11 = 88$.
Subtracting 88 gives 0.
The answer is 18.

Written Exercise

258. Find the answer:

1. 132 ×11	2. 234 ×11	3. 456 ×11	4. 567 ×11	5. 678 ×11
6. 231 ×11	7. 342 ×11	8. 564 ×11	9. 675 ×11	10. 786 ×11
11. 109 ×11	12. 207 ×11	13. 308 ×11	14. 406 ×11	15. 503 ×11
16. 120 ×11	17. 320 ×11	18. 340 ×11	19. 560 ×11	20. 600 ×11

21. $\begin{array}{r} 120 \\ \times 11 \\ \hline \end{array}$ 22. $\begin{array}{r} 1,030 \\ \times 11 \\ \hline \end{array}$ 23. $\begin{array}{r} 3,201 \\ \times 11 \\ \hline \end{array}$ 24. $\begin{array}{r} 1,079 \\ \times 11 \\ \hline \end{array}$ 25. $\begin{array}{r} 876 \\ \times 11 \\ \hline \end{array}$

26. $11\overline{)132}$ 27. $11\overline{)143}$ 28. $11\overline{)154}$

29. $11\overline{)165}$ 30. $11\overline{)110}$ 31. $11\overline{)1100}$

32. $11\overline{)1,210}$ 33. $11\overline{)1,320}$ 34. $11\overline{)1,430}$

Written Problems

259. 1. A freight train consists of 11 cars. Each car carries 196 bbl. of lime. How many barrels are there in the 11 cars?

2. Eleven cakes of ice weigh 120 lb. each. What is the weight of the 11 cakes of ice?

3. The distance from *A* to *B* is 11 times the distance from *A* to *C*. If the distance from *A* to *C* is 21 mi., how far is it from *A* to *B*? Illustrate by diagram.

4. A hardware dealer sells the following bill of goods:

11 saws @ $1 each.
11 hammers @ $1.10 each.

Find the amount of the bill.

5. A boy saved 11¢ each week for 12 weeks. How much money did he save?

6. A florist sold 11 plants for $33. What was the average price received for each plant?

7. A motor boat travels 99 mi. in 3 hr. How far does the motor boat travel in 1 hr.?

8. If the motor boat travels 11 times as fast as a rowboat, how far does the rowboat travel in 1 hr.?

9. An automobile travels 220 mi. in 11 hr. If no allowance is made for stops, what is the average distance traveled in an hour.?

10. An express train travels 660 mi. in 11 hr. What is the average rate of speed per hour?

Problem Making

260. Make problems using some of the following combinations:

1. 11 bbl. at $4.	2. 12 lb. at 11¢.
3. 11 bu. at 60¢.	4. 27 yd. at 11¢.

Multiplication and Division by Twelve — Oral Exercise

261. Give the products:

1. 2×12	2. 3×12	3. 4×12
4. 5×12	5. 6×12	6. 7×12
7. 8×12	8. 9×12	9. 10×12

262. Make and memorize the multiplication table of twelve.

Count:

1. By 12's to 60, beginning with 12.
2. By 12's to 96, beginning with 12.
3. By 12's to 144, beginning with 12.

4. By 12's to 144, beginning with 60.
5. By 12's to 144, beginning with 96

263. Give the quotients:

1. $12\overline{)24}$ 2. $12\overline{)36}$ 3. $12\overline{)60}$ 4. $12\overline{)72}$

5. $12\overline{)96}$ 6. $12\overline{)144}$ 7. $12\overline{)120}$ 8. $12\overline{)132}$

Oral Problems

264. 1. How many inches in 1 ft.? In 2 ft.? In 5 ft.?

2. How many feet in 72 in.? In 60 in.? In 84 in.? In 36 in.?

3. How many apples in 1 doz. apples? In 2 doz.? In 4 doz.?

4. How many dozen stamps in 96 stamps? In 84 stamps? In 48 stamps?

5. If twelve pencils cost 24 ¢, what part of 24 ¢ is the cost of one pencil? What is the cost of one pencil?

6. Twelve tops cost 60 ¢. How many tops can be bought for 30 ¢?

7. A dozen bananas cost 36 ¢. What is the cost of one banana?

8. A 12-story building is 144 ft. high. What is the average height of each story?

9. If the cellar of this building is 12 ft. deep, what is the distance from the top of the building to the cellar?

10. A box contains 1 doz. cards of buttons. There are 1 doz. buttons on each card. How many buttons in the box?

Drill Device — The Circle

265. Multiply each number on the circle by 12. Vary the order.

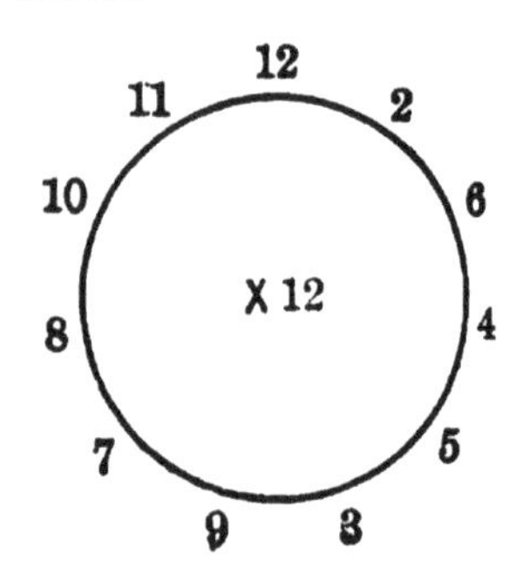

Make a drill device for division by 12. Use on the device numbers that are exactly divisible by 12.

Drill to obtain quickly the quotients.

Written Exercises

266. Find the answer:

1. 122 × 12	2. 234 × 12	3. 345 × 12	4. 456 × 12
5. 577 × 12	6. 678 × 12	7. 765 × 12	8. 465 × 12
9. 209 × 12	10. 307 × 12	11. 408 × 12	12. 506 × 12
13. 220 × 12	14. 340 × 12	15. 450 × 12	16. 650 × 12

17. $\begin{array}{r} 2{,}104 \\ \times 12 \\ \hline \end{array}$ 18. $\begin{array}{r} 3{,}010 \\ \times 12 \\ \hline \end{array}$ 19. $\begin{array}{r} 2{,}031 \\ \times 12 \\ \hline \end{array}$ 20. $\begin{array}{r} 1{,}709 \\ \times 12 \\ \hline \end{array}$

21. $12\overline{)144}$ 22. $12\overline{)132}$ 23. $12\overline{)120}$ 24. $12\overline{)108}$

25. $12\overline{)240}$ 26. $12\overline{)360}$ 27. $12\overline{)480}$ 28. $12\overline{)96}$

29. $12\overline{)300}$ 30. $12\overline{)420}$ 31. $12\overline{)180}$ 32. $12\overline{)600}$

Written Problems

267. 1. Find the cost of 12 doz. inkwells at 39 ¢ a doz.

2. A lady bought 26 yd. of ribbon at 12 ¢ a yard. She gave in payment a five-dollar bill. How much change should she receive?

3. There are 40 pupils in each class in a school of 12 rooms. How many pupils in the school?

4. A sewing machine costs $ 65. Find the cost of 12 sewing machines.

5. How many pounds of butter are there in 12 tubs, if each tub contains 45 lb. of butter?

6. A building contains 12 apartments. The average rent for each apartment is $ 35 a month. What is the yearly rental received from this building?

7. A train is made up of 12 box cars, each car containing 360 boxes of oranges. How many boxes of oranges in the 12 cars?

8. A furniture dealer sold 108 tables at an average price of $12 each. He received in part payment a check for $1,000. What is the balance still due him?

9. A printing press prints 2,364 papers in an hour. How many papers will it print in 12 hr. at the same rate?

10. A monthly commutation ticket costs $6.30. What is cost of commutation for a year?

11. There are 480 pupils in a school of 12 rooms. What is the average number of pupils in each room?

12. A plot of ground is 192 ft. long. Its length is divided so as to make 12 lots of equal size. What is the width of each lot?

13. A monitor distributed 12 sheets of paper to each pupil in the class. If he distributed 422 sheets, how many pupils are there in the class?

14. I put 12 doz. pins into each of 12 boxes. If I had 2,000 pins, how many pins are left?

15. How many full boxes can be made from the pins left over?

16. A bar of lead weighs 12 lb. How many bars of lead will be required to make a weight of 168 lb.?

FACTORS

268. 1. What two numbers multiplied together give 9 as a product? 3 and 3 are factors of 9.

2. What two numbers multiplied together give 14 as a product? 2 and 7 are factors of 14.

3. What two numbers multiplied together give 21 as a product? 3 and 7 are factors of 21.

4. What two numbers multiplied together give 15 as a product? 3 and 5 are factors of 15.

5. What two numbers multiplied together give 33 as a product? 3 and 11 are factors of 33.

6. What 3 numbers multiplied together give 8 as a product? $2 \times 2 \times 2$ are factors of 8.

7. What 3 numbers multiplied together give 12 as a product? $2 \times 2 \times 3$ are factors of 12.

8. What 3 numbers multiplied together give 20 as a product? $2 \times 2 \times 5$ are factors of 20.

Oral Exercises

269. Find the missing factors:

Two is one factor of each of the following numbers. What is the other?

1. 14	2. 22	3. 24	4. 32	5. 36
6. 42	7. 16	8. 18	9. 48	10. 40

Three is one factor. What is the other?

1. 24	2. 27	3. 45	4. 48	5. 36
6. 18	7. 21	8. 9	9. 30	10. 33

Four is one factor. What is the other?

1. 16	2. 8	3. 24	4. 36	5. 40
6. 48	7. 32	8. 28	9. 44	10. 12

270. Give one pair of factors of each of the following numbers:

1. 56	2. 63	3. 64	4. 54	5. 49
6. 48	7. 24	8. 42	9. 44	10. 39
11. 72	12. 77	13. 66	14. 81	15. 84
16. 36	17. 25	18. 35	19. 55	20. 65
21. 75	22. 85	23. 15	24. 45	25. 5
26. 30	27. 40	28. 50	29. 60	30. 70

Review Device — The Wheel

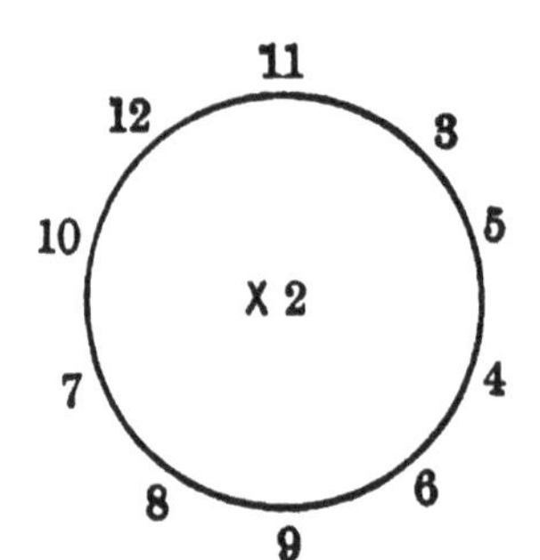

271. Multiply the numbers on the rim by the number in the center. Change the number in the center, using 4, 5, 6, 7, 8, and 9.

Divide the numbers on the rim by the number in the center. Change the numbers on the rim to numbers exactly divisible by 3, and divide by 3. Do the same for 4, 5, 6, etc.

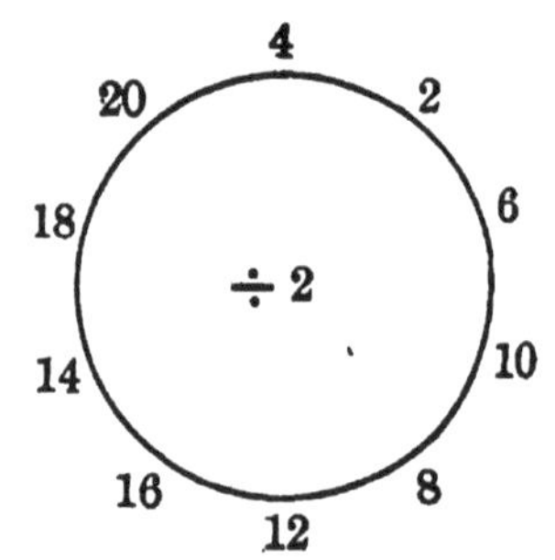

Frequent drills should be given upon the relations that are found most difficult. Drill devices containing these difficult relations should be used.

The various forms in which it is possible to express the relations should be drilled, *e.g.*, $8 \times 7 = 56$. Variations of this relation are $7 \times 8 = ?$ How many 8's in 56? How many 7's in 56? $8 \times ? = 56$. $7 \times ? = 56$. $? \times ? = 56$.

Multiplication — Written Exercises

272. Multiply:

1.	24	36	47	55	63	72
	9	8	7	6	5	4
2.	112	256	432	524	675	916
	4	5	6	7	8	9
3.	35	75	95	85	105	225
	10	10	10	10	10	10
4.	121	141	181	221	351	271
	10	10	10	10	10	10
5.	106	203	407	504	608	100
	10	10	10	10	10	10
6.	20	40	50	60	70	80
	10	10	10	10	10	10
7.	220	350	460	570	680	701
	10	10	10	10	10	10

8.	200	300	400	500	600	1000
	10	10	10	10	10	10

9.	201	300	29	99	88	77
	9	5	7	9	8	7

10.	999	888	777	666	555	987
	8	9	6	7	4	9

Division — Oral Exercises

273. Find the answers:

1.	2)20	2)40	2)60	2)80	2)90	2)70
2.	3)21	3)30	3)27	3)24	3)36	3)45
3.	4)24	4)32	4)44	4)52	4)64	4)56
4.	5)35	5)40	5)45	5)50	5)55	5)60
5.	6)48	6)36	6)30	6)42	6)54	6)72
6.	7)42	7)35	7)56	7)49	7)63	7)77
7.	8)32	8)40	8)56	8)64	8)72	8)48
8.	9)36	9)54	9)45	9)63	9)81	9)72
9.	10)40	10)50	10)60	10)70	10)80	10)90

274. **Written Exercise**

1. $7\overline{)147}$	2. $8\overline{)328}$	3. $9\overline{)549}$	4. $6\overline{)486}$
5. $5\overline{)745}$	6. $8\overline{)448}$	7. $7\overline{)714}$	8. $9\overline{)954}$
9. $5\overline{)455}$	10. $6\overline{)396}$	11. $5\overline{)205}$	12. $6\overline{)306}$
13. $7\overline{)707}$	14. $8\overline{)504}$	15. $9\overline{)801}$	16. $9\overline{)810}$
17. $8\overline{)640}$	18. $7\overline{)350}$	19. $6\overline{)540}$	20. $5\overline{)450}$

Multiplication — Written Exercise

275. 1. Mr. Thomas sells 5 city lots for $625 each. How much does he receive for the lots?

2. Find the cost of 4 pianos at $375 each.

3. One farmer sells 524 sheep at $4 each. Another farmer receives $5 each for 419 sheep. How much does each farmer receive for his sheep?

4. If 1 doz. eggs cost 48¢, what is the cost of 10 doz.?

5. A dealer sold 7 automobiles for $750 each. How much did he receive?

Division — Written Exercise

276. 1. A picnic party of 8 spent $16 for expenses. What was the share of each?

2. How many pounds of apricots, at 9¢ a lb., can be bought for 99¢?

3. If 8 tons of hay cost $ 96, what is the price of 1 ton?

4. The distance from *A* to *B* is 45 mi. The distance from *A* to *C* is one third as great. What is the distance from *A* to *C*?

5. A farm contains 280 acres. How many acres in a farm one seventh as large?

277. Multiplication — Written Exercises

1. 1,495 6	2. 2,306 7	3. 1,087 8	4. 4,261 9
5. 5,941 9	6. 3,062 8	7. 8,071 7	8. 2,614 6
9. 45,327 8	10. 63,471 9	11. 2,357 10	12. 3,752 10
13. 315 11	14. 409 11	15. 478 11	16. 389 11
17. 1,930 11	18. 2,031 11	19. 208 12	20. 316 12
21. 789 12	22. 1,021 12	23. 2,760 12	24. 3,781 12

278. Review — Four Operations

1. $130 \div 10 + 1$
2. $180 \div 10 + 1$
3. $180 \div 10 - 1$
4. $200 \div 10 + 9$
5. $100 \div 10 + 1$
6. $100 \times 1 + 3$
7. $3 \times 10 - 6$
8. $7 \times 10 - 8$
9. $17 \times 10 - 10$
10. $7 \times 9 - 10$
11. $6 \times 7 - 10$
12. $8 \times 9 + 10$
13. $2 + 9 \times 6$
14. $2 + 7 \times 8$
15. $3 + 6 \times 7$
16. $12 \times 3 - 10$
17. $12 \times 4 - 20$

1 . $12 \times 5 - 10$

19. $5 \times 5 + 20$
20. $4 \times 4 + 10$
21. $3 \times 3 + 30$
22. $7 \times 7 - 10$
23. $6 \times 6 + 10$
24. $10 \times 60 - 100$
25. $10 \times 40 - 100$
26. $10 \times 30 + 200$
27. $10 \times 17 - 100$
28. $10 \times 3 + 20$
29. $180 \div 10 + 2$
30. $121 \div 11 + 1$
31. $99 \div 11 + 11$
32. $88 \div 11 + 20$
33. $66 \div 11 + 11$
34. $77 \div 11 + 10$
35. $55 \div 5 - 10$
36. $44 \div 4 - 10$

Perform multiplication or division before the addition and subtraction.

Problem Making

279. Make problems, using some of the following numbers:

1. 14 lb. at 7¢.
2. 27 yd. at 9¢.
3. 6 lb. at 36¢.
4. 36 lb. at 6¢.
5. 35 yd. at 8¢.
6. 42 bu. at 40¢.

7. Make problems about the things that can be bought at the school lunch counter.

8. Make problems from the articles mentioned in the grocery list.

Grocery List

Milk, 9¢ per qt.
Molasses, 10¢ per pt.
Cider, 11¢ per qt.
Kerosene Oil, 10¢ half gal.
Spinach, 8¢ per qt.
String Beans, 9¢ per qt.
Peas, 10¢ per qt.
Tomatoes, 11¢ per qt.
Apples, 12¢ per qt.

Written Exercises

280. Add by line and by column:

1.
$56 + 71 + 88 = ?$
$38 + 84 + 46 = ?$
$21 + 37 + 83 = ?$
$? + ? + ? = ?$

2.
$18 + 96 + 85 =$
$59 + 96 + 95 =$
$50 + 80 + 10 =$
$? + ? + ? = ?$

3.
$178 + 943 + 754 = ?$
$347 + 42 + 202 = ?$
$858 + 40 + 10 = ?$
$507 + 969 + 959 = ?$
$? + ? + ? = ?$

4.
$183 + 432 + 542 =$
$748 + 209 + 502 =$
$585 + 80 + 4 =$
$79 + 696 + 559 =$
$? + ? + ? = ?$

5. Subtract 1,234 from 3,702; subtract 1,234 from the resulting remainder, and so on. If you subtract correctly, the last remainder will be zero.

6. Proceed in the same way with 4,567 and 13,701.

7. Proceed in the same way with 7,890 and 31,560.

Written Problems

281. 1. One man spends $7 a month and another $5 a month. How much do they both spend in a month? In a year?

2. A man had $125 in the bank. To this amount he added his savings of $12 a month for 7 months. How much was his bank account at the end of the 7 months?

3. A farmer sold 10 cows at $36 each. With this money he bought 8 pigs at $12 each. How much money had he left?

4. A man rode for 5 hr at the rate of 12 mi. an hour. The following day he rode for 7 hr. at the same rate. How far did he ride in the 2 da.?

5. I bought goods from a merchant as follows:

March $186.75, April 65.70, May 240.25.

On the first of June I paid the merchant $300 on account. How much do I still owe?

6. From the school garden plot, a boy sold the following:

12 heads of lettuce	10¢ per head
24 bunches of radishes	3 bunches for 5¢
10 qt. of peas	9¢ per quart
5 bunches of beets	5¢ per bunch

How much did he realize from the sale?

7. A real estate agent received $ 450 for a year's rent, of which he spent $ 75 for repairs and $ 50 for taxes. How much remained?

8. A man leased a house for $ 240 for one year. How much was the rent per month?

9. A man's estate of $ 4500 was left to his widow and three sons. The widow received $ 1500 and the remainder was equally divided among the sons. What was the share of each son?

DIVISION

Division Involving Remainders — Written Exercise

282. Divide:

1. 273 by 2.

PROCESS

```
   136½
 ______
2)273
  2
  -
   7
   6
  --
   13
   12
   --
    1 Remainder
```

FULL EXPLANATION. — 2 ÷ 2 = 1. Write the 1 above the line over the 2.

Multiply the divisor by the 1 in the quotient.

Write the product, 2, under the 2 in the dividend and subtract.

Bring down the 7, the next digit in the dividend.

2 is contained in 7 three times. Write the 3 above the line over the 7.

Multiply and subtract. Write the remainder 1 below the line and bring down the 3 from the dividend.

2 is contained in 13 six times. Write the 6 above the line over the 3.

Multiply and subtract.

There are no other figures in the dividend to be brought down.

The quotient is 136 and the remainder is 1. The result of the division may also be written $136\frac{1}{2}$.

This process is called *long division.*

2. 375 by 6.

PROCESS

```
    62
6) 375
   36
    15
    12
     3 Remainder
```

BRIEF EXPLANATION. — 6 is not contained in the first figure in the dividend. 6 is contained in 37 six times.

Write the 6 above 7, the second figure in the dividend.

Multiply and subtract. Bring down the next figure in the dividend.

6 is contained in 15 twice. Write the 2 above 5, the third figure in the dividend.

Multiply and subtract.

The quotient is 62, the remainder 3, and the answer $62\frac{3}{6}$ or $62\frac{1}{2}$.

It may be helpful to fill in the vacant places to the left of the quotient with small crosses.

Written Exercises

283. Divide and explain the process:

1. $2\overline{)901}$ 2. $3\overline{)902}$ 3. $3\overline{)100}$ 4. $3\overline{)731}$

5. $3\overline{)98}$ 6. $2\overline{)99}$ 7. $4\overline{)75}$ 8. $6\overline{)75}$

9. $4\overline{)100}$	10. $6\overline{)100}$	11. $3\overline{)245}$	12. $4\overline{)241}$
13. $6\overline{)250}$	14. $7\overline{)238}$	15. $8\overline{)120}$	16. $9\overline{)120}$
17. $7\overline{)900}$	18. $8\overline{)910}$	19. $9\overline{)910}$	20. $7\overline{)100}$

Written Problems

284. 1. Reduce 74 pk. to bushels.

PROCESS

$$\begin{array}{r} 18 \\ 4\overline{)74} \\ \underline{4} \\ 34 \\ \underline{32} \\ 2 \end{array}$$

EXPLANATION. — 4 pk. make 1 bu. To reduce 74 pk. to bushels, divide 74 by 4. The answer is 18 bu. 2 pk. or $18\frac{1}{2}$ bu.

2. Reduce 94 ft. to yards and a fraction of a yard. Write the answer also in yards and feet.

3. How many quarts in 95 pt.?

4. How many gallons in 47 qt.?

5. How many yards in 140 ft.?

6. How many weeks in 152 da.?

7. How many feet in 60 in.?

8. How many dimes in 98¢?

9. How many pecks in 142 qt.?

10. How many years in 135 mo.?

11. A boy ran 100-yd. dash in 11 sec. Is it more than 9 yd. per second? Is it more than 10? What is the average number of yards per second?

12. An athlete put the shot 37 ft. How many yards in this distance?

13. A boy made a broad jump, 22 ft. Is this distance greater than 9 yd.?

14. In a mile there are 1760 yd. How many yards in one fourth mile?

Tell how to reduce

inches to feet	pints to quarts
quarts to pecks	quarts to gallons
pecks to bushels	days to weeks

months to years

Tell how to change smaller units to larger units of measure.

Written Exercises

285. Divide:

1. 865 by 7.

LONG DIVISION PROCESS	SHORT DIVISION PROCESS
$$\begin{array}{r} 123 \\ 7\overline{)865} \\ \underline{7} \\ 16 \\ \underline{14} \\ 25 \\ \underline{21} \\ 4 \end{array}$$ *Remainder.*	$$\begin{array}{l} \overline{123 \text{ and } 4 \text{ remainder}} \\ 7)865 \end{array}$$

EXPLANATION. — Long division was explained in Ex. 284.

Short division: $8 \div 7 = 1$, and 1 over. Write the 1 above the 8. Subtracting 1×7 from 8 leaves 1.

$16 \div 7 = 2$, and 2 over. Write

the 2 above the 6. Subtracting 2×7 from 16 leaves 2.

$25 \div 7 = 3$, and 4 over. Write the 3 above the 5. The answer is $123\frac{4}{7}$.

Compare the process in long division with the process in short division. Explain the difference.

Written Exercises — Short Division

286. 1. 2)43 2. 2)95 3. 3)31 4. 3)76

5. 3)87 6. 4)66 7. 4)74 8. 5)51

9. 5)84 10. 4)91 11. 6)71 12. 6)95

13. 7)83 14. 7)99 15. 7)92 16. 8)98

17. 8)89 18. 8)79 19. 9)76 20. 9)98

21. 3)105 22. 4)105 23. 5)106 24. 6)103

25. 7)107 26. 7)201 27. 8)205 28. 9)202

29. 8)311 30. 7)305 31. 4)300 32. 6)400

33. 7)500 34. 8)600 35. 9)700 36. 7)906

Written Problems

287. 1. Reduce 835 ft. to yards and fraction of a yard. Write the answer in yards and feet.

2. How many weeks in 365 da.?

3. How many bushels in 315 pk.?

4. How many gallons in 275 pt.?

5. How many boxes containing one half dozen each can a grocer fill from a shipment of 445 eggs?

6. How many boxes containing 12 peaches each can be filled from a basket containing 200 peaches?

7. The length of a classroom is 25 ft. 6 in. How many inches long is it?

8. The length of a study hall is 42 ft. 9 in. What is its length in inches?

9. What is the difference in length between a classroom that is 24 ft. 8 in. long and one that is 32 ft. 4 in. long? Give the answer in inches.

10. In a pole vault a boy reaches the height of 8 ft. 6 in. The record for the school is 10 ft. Find the difference in inches.

11. A jumper cleared 6 ft. 4 in. How much more is this than 2 yd.? How much greater is this than the school record of 5 ft.?

12. The champion broad jump for one year was 24 ft. 7 in. How much greater is this than the record of the previous year of 20 ft.?

Estimating Weights

288. 1. Compare the weight of this arithmetic with the $\frac{1}{2}$-lb. weight. Which is the heavier?

2. Compare the weight of two books with the pound weight.

3. How many pounds do you weigh?

4. Give the weights in pounds of the first two pupils in your row.

5. Give the total weight of 4 pupils in your class.

Avoirdupois Weight — Oral

289 1. How many ounces in 1 lb.?

2. How many ounces in $\frac{1}{2}$ lb.?

3. How many ounces in $\frac{1}{4}$ lb.?

4. What part of a pound is 8 oz.?

5. What part of a pound is 4 oz.?

6. What weight is used in weighing coal?

7. How many pounds in a ton?

8. How many pounds in $\frac{1}{2}$ T.? In $\frac{1}{4}$ T.?

AVOIRDUPOIS WEIGHT

16 ounces (oz.) = 1 pound (lb.)
100 pounds = 1 hundred weight (cwt.)
20 hundred weight or 2000 pounds = 1 ton (T.)

Oral Problems

290. 1. Rice costs 4¢ a lb. How many pounds can be bought for 48¢?

2. How many ounces of rice can be bought for 2¢ if it cost 4¢ per lb.?

3. A half pound of coffee cost 28¢. How much coffee can be bought for 84¢?

4. How many pounds of sugar at 6¢ a pound will pay for 8 qt. of strawberries at 9¢ a quart?

5. If 25 lb. of flour cost $ 1, how much will 100 lb. cost?

6. A boy bought the following articles at the grocery store:

5 lb. sugar at 6¢ per pound.
2 lb. of butter at 30¢ per pound.

He gave the storekeeper $ 1 bill. How much change did he receive?

7. If a pound of tea costs 60¢, how many pounds can be bought for $ 2?

8. If $\frac{1}{2}$ lb. of cheese costs 10¢, how much must be paid for 2 lb.?

9. A liveryman bought 350 lb. of pressed hay for $ 3.50. How much did it cost him per lb.?

10. A farmer seeded 5 A. with bluegrass seed, using 25 lb. to the acre. How many pounds of seed did he use?

11. A farmer has 100 lb. of timothy seed. If it requires 20 lb. of seed for an acre, how many acres can he seed?

12. A seed dealer has 600 lb. of alfalfa seed which he puts up in sacks holding 30 lb. How many sacks does he fill?

13. How many hundredweight in 2 T. of coal?

14. How many pounds in 2 T. of coal?

15. How many pounds in $\frac{1}{2}$ T. of coal?

Oral Exercises

291. Give the answer:

1. 11 lb. $\times 8$
2. 12 lb. $\times 8$
3. 12 lb. $\times 9$
4. 12 lb. $\times 10$
5. 12 lb. $\times 7$
6. 12 lb. $\times 6$
7. 8 lb. $\times 11$
8. 8 lb. $\times 12$
9. 9 lb. $\times 12$
10. 10 lb. $\times 12$
11. $4\overline{)12}$ lb.
12. $4\overline{)16}$ lb.
13. $7\overline{)28}$ lb.
14. $11\overline{)121}$ oz.
15. $12\overline{)144}$ oz.
16. $\frac{1}{2}$ of 24 lb.
17. $\frac{1}{4}$ of 36 lb.
18. $\frac{1}{3}$ of 33 lb.
19. 14 T. $\times 5$
20. 17 T. $\times 2$
21. 11 T. $\times 7$
22. 12 T. $\times 11$
23. 11 T. $\times 12$

Written Exercises

292. Give the answer:

1. 101 lb. $\times 12$
2. 201 lb. $\times 12$
3. 202 lb. $\times 12$
4. 303 lb. $\times 12$
5. 404 lb. $\times 12$
6. 707 lb. $\times 12$
7. 808 oz. $\times 12$
8. 505 T. $\times 12$
9. 607 T. $\times 12$
10. 509 lb. $\times 12$
11. 509 $\times 11$
12. 589 $\times 11$
13. $5\overline{)105}$ lb.
14. $6\overline{)612}$ oz.
15. $7\overline{)840}$ T.
16. $9\overline{)198}$ lb.
17. $10\overline{)730}$ oz.
18. $9\overline{)207}$

The Months — Drill Device

293. "Thirty days hath September," etc.

To tell the number of days in a month, touch the knuckles and hollows in the closed hand as shown in the picture, naming the months in their order. When you have named *July* begin again with the first knuckle and name the remaining months. The knuckles represent the months containing 31 days.

How many days in

1. February of the present year?
2. The present year?
3. A leap year?
4. June and July?
5. September and November?
6. July and August?
7. April, June, and September?
8. January, March, and August?

Written Exercises

294. 1. A farmer loads his wagon with 9 sacks of potatoes, each weighing 125 lb. What is the total weight of the load of potatoes? How many pounds does it lack to make 1 T.?

2. If a limb of a tree can carry 1000 lb., how many boys weighing 100 lb. each will it support?

3. What is the weight of 7 hams, each weighing 15 lb.?

4. A barrel of flour weighs 196 lb. What is the weight of 7 bbl.?

5. If a barrel weighs 13 lb. when empty, and 319 lb. when filled with sugar, what is the weight of the sugar?

6. If 6 boys weigh 486 lb., what is the average weight of 1 boy?

7. A boy carries 8 pails of water, each weighing 65 lb. How many pounds did he carry?

8. From a ton of coal a man sold 6 sacks of coal each weighing 123 lb. How much coal is left?

9. A well-balanced daily ration for dairy cattle is made up of 20 lb. of hay, 4 lb. of oats, 4 lb. of corn for each cow. What is the total number of pounds of fodder required for one day for nine cows?

10. Give the total number of pounds of fodder for 1 cow for 1 wk. For 4 wk.

11. Give the total number of pounds of fodder required for 12 cows for 1 wk. For 4 wk.

12. A load of hay and the wagon weigh together 2950 lb. The wagon weighs 1145 lb. What is the weight of the hay? How much does the hay lack of weighing a ton?

Multipliers 10, 20, etc.

295. 1. Multiply 18 by 20.

PROCESS

```
 18
 20
---
360
```

EXPLANATION. — Arrange the numbers, placing the 2 in the multiplier under the ones' place in the multiplicand.

20 = 2 tens.

20 × 18 is the same as

2 tens × 18 = 36 tens or 360.

A short method of multiplying by 20 is to multiply by 2 and annex the 0 to the product.

2. Multiply 95 by 30.

PROCESS

```
   95
   30
-----
2,850
```

EXPLANATION. — Arrange the numbers. Write the explanation.

Give the short method of multiplying by 30.

3. Multiply 264 by 400.

PROCESS

```
    264
    400
-------
105,600
```

EXPLANATION. — Arrange the numbers, placing the 4 in the multiplier under the ones' place in the multiplicand. Write the explanation.

Give the short method of multiplying by numbers that end in one cipher or two ciphers.

Written Exercises

296. Multiply:

1.	2.	3.	4.	5.
64	75	67	126	231
30	40	50	80	90

6.	7.	8.	9.	10.
32	47	59	73	65
100	200	300	500	700

Divisors 10, 20, etc.

297. Divide:

1. 370 by 10.

PROCESS

$$\begin{array}{r} 37 \\ 1\not{0})\overline{37\not{0}} \end{array}$$

EXPLANATION. — 370 = 37 tens.
1 ten is contained in 37 tens 37 times.
37 is the quotient.

A short method of dividing a number that ends with a **0** by **10** is to strike out or cancel the **0** at the right of the divisor and the dividend, and divide the resulting numbers.

2. 5,760 by 20.

PROCESS

$$\begin{array}{r} 288 \\ 2\not{0})\overline{5,76\not{0}} \end{array}$$

Write the explanation.

3. 98,460 by 50.

PROCESS

$$\begin{array}{r} 1969 \\ 5\not{0})\overline{98,45\not{0}} \end{array}$$

Write the explanation.

298.

1. $10\overline{)480}$ 2. $10\overline{)1,850}$ 3. $10\overline{)6,230}$ 4. $10\overline{)3,470}$

5. $10\overline{)5,460}$ 6. $10\overline{)1,050}$ 7. $10\overline{)3,070}$ 8. $10\overline{)8,090}$

9. $10\overline{)7,060}$ 10. $10\overline{)6,070}$ 11. $20\overline{)780}$ 12. $20\overline{)8,060}$

13. $30\overline{)6,390}$ 14. $40\overline{)4,640}$ 15. $50\overline{)7,050}$ 16. $60\overline{)7,260}$

17. $70\overline{)4,970}$ 18. $80\overline{)5,680}$ 19. $9\overline{)8,190}$ 20. $9\overline{)1,890}$

Division

299. Divide:

1. 5630 by 20.

PROCESS

$$\begin{array}{r} 281\tfrac{1}{2} \\ 2\not{0}\overline{)5,63\not{0}} \end{array}$$

Write the explanation.

2. 30,450 by 70.

PROCESS

$$\begin{array}{r} 435\tfrac{1}{7} \\ 7\not{0}\overline{)30,46\not{0}} \end{array}$$

Write the explanation.

Written Exercises

300. Divide:

1. $20\overline{)4,350}$ 2. $40\overline{)3,270}$ 3. $70\overline{)56,740}$

4. $90\overline{)382,560}$ 5. $50\overline{)23,540}$ 6. $30\overline{)32,730}$

7. $80\overline{)467,540}$ 8. $60\overline{)823,560}$ 9. $50\overline{)35,240}$

10. $70\overline{)73,230}$ 11. $80\overline{)765,440}$ 12. $90\overline{)563,280}$

Written Problems

301. 1. What is the cost of 100 T. of hay at $16 a ton?

2. A bushel of wheat weighs 60 lb. What is the weight of a load of wheat which contains 54 bu.?

3. A train travels at the rate of 50 mi. an hour. How far does it travel in 12 hr.? In 24 hr.?

4. What is the cost of 20 lb. of salmon at 13¢ a pound?

5. If trout costs 20¢ a pound, what will 11 lb. cost?

6. Halibut sells for 15¢ a pound. What must you pay for 30 lb.?

BILLS AND RECEIPTS

302. 1. A photographer sold a kodak for $15, a carrying case for $1.25, and 6 rolls of films for 35¢ each. What was the amount of the sale?

2. A dealer sold a suit case for $7.50, a trunk for $12 and a harness for $16.50. How much did he receive in payment?

3. A merchant buys tea for 50¢ a pound and sells it for 60¢. What is his profit on 48 lb.?

4. A storekeeper buys a barrel of flour (containing 196 lb.) at $3.75. He sells the flour by the

pound at 4¢ a pound. How much does he make on the sale?

5. A farmer sold 15 doz. eggs at 30¢ a dozen. How much did he receive?

6. If a farmer buys $18.50 worth of groceries and sells the grocer $8.75 worth of butter, how much money does he have to pay the grocer?

7. A grocer buys prunes for $8.75, cheese for $11.50, a barrel of sugar for $6.75; if he gives the salesman $30, how much change does he receive?

8. In a garden bed, 48 yd. long, strawberry plants are planted 1 ft. apart, in rows running lengthwise, the outer plants being 1 ft. from the boundary. How many plants are there in a row? How many plants are there in three rows?

9. A car contains 150 barrels of apples. If each barrel contains 3 bushels, how many bushels of apples are there in the car?

10. A farmer pays $175 for a horse, $67 for a harness and $225 for a buggy. How much does he pay altogether?

11. A gardener sets out plants 6 in a row. How many rows can he plant with 225 plants? How many plants are left over?

Bills and Receipts

303. 1. Mrs. Frank Robinson, 120 Michigan Avenue, Chicago, visited the store of Marshall Field & Company of Chicago on December 1, 1913. She ordered the following goods:

2 pr. gloves at	$2.50
4 ties at	1.00
3 boxes men's hdkfs. at	1.50
1 lace collar	3.75

The firm of Marshall Field & Company sent Mrs. Robinson the following bill on January 2, 1914:

CHICAGO, JAN. 2, 1914.

MRS. FRANK ROBINSON,
120 Michigan Ave.,
Bought of MARSHALL FIELD & CO.

1918					
Dec.	1	2	pr. gloves @ $2.50	5	00
		4	ties at $1.00	4	00
		3	boxes hdkfs. @ $1.50	4	50
		1	lace collar @ $3.75	3	75
				$17	25

(*a*) What is the date on which the goods were bought? Where is it written on this bill?

(*b*) On what date did Marshall Field & Co. send the bill? Where is it written on the bill?

(*c*) Find the column in which is written the number of articles purchased.

(*d*) Where is the cost of each item written?

(*e*) Where is the total amount of the bill written?

2. Make a bill for some dry goods bought by yourself from the local dealer.

3. Make a bill for some groceries bought from the local grocer.

4. Make a bill for some meat bought from the local butcher.

5. Mrs. Frances Brown bought from her grocer, Charles Bedell, on December 23, 1913, the following articles:

2 boxes of raisins at 20¢ a box.
2 boxes of currants at 18¢ a box.
3 qt. cranberries at 15¢ a qt.
1 large bunch of celery at 40¢.

(*a*) What is the amount of each item?
(*b*) What is the total amount of the bill?

6. Mrs. Frances Brown paid Charles Bedell for the goods purchased December 23, and received a receipted bill.

NEW YORK CITY, DEC. 23, 1913.

MRS. FRANCES BROWN,
25 Manhattan Ave.,

Bought of CHARLES BEDELL

Dec.	23	2	boxes of raisins @ 20 ct.		40
		2	boxes of currants @ 18 ct.		36
		3	qt. of cranberries @ 15 ct.		45
		1	bunch of celery @ 40 ct.		40
				$1	61
			Received payment CHARLES BEDELL		

Other forms of receipting bills are:

Received Payment
CHARLES BEDELL,
Per A. HULL.

December 23, 1913.

Paid
CHARLES BEDELL.
A. HULL.

(*a*) Where is the receipt written on the bill?

(*b*) Who receipted the bill?

(*c*) In what other ways might the bill have been receipted?

7. Fill in the blank spaces in the following bill:

JANUARY 2, 1914

____

____

____

1913					
Dec.	10	5	lb. powdered sugar @ 6 ct.		
	12	½	doz. oranges @ 5 ct. each.		
	15	3	grapefruit @ 3 for 25 ct.		
			________________________	___	___
			________________________		

(*a*) On what day was the first item on the bill purchased?

(*b*) On what day was the bill made out?

8. Suppose you are a clerk in the store of Hemenway & Company. Make out a bill of goods bought by one of your schoolmates and receipt it.

9. John Jones works on Mr. Henry Stone's farm. He began work on April 1, 1914, at $ 2 a day. On April 30 he sent a bill for services to to Mr. Stone.

NUTLEY, N. J., APRIL 30, 1914.

MR. HENRY STONE
To John Jones, Dr.

1914 April	1 to 6	To 6 days' service @ $2 a day		
	8 to 12	To 5 days' service @ $2 a day		
	16 to 19	To 4 days' service @ $2 a day		
	27 to 29	To 3 days' service @ $2 a day		

(*a*) Find the total amount of Mr. Jones' bill.
(*b*) Receipt this bill for John Jones.

10. Make out a receipted bill for services rendered by yourself.

LONG DIVISION

304. Divide:

1. 987 by 8.

PROCESS

$$\begin{array}{r} 123\frac{3}{8} \\ 8\overline{)987} \\ \underline{8} \\ 18 \\ \underline{16} \\ 27 \\ \underline{24} \\ 3 \end{array}$$

EXPLANATION. — 987 = 9 hundreds + 8 tens + 7 ones.

$$8\overline{)9 \text{ hundreds} + {}^{1}8 \text{ tens} + {}^{2}7 \text{ ones}}\quad \text{with quotient } 1 \text{ hundred} + 2 \text{ tens} + 3 \text{ ones}, \; 3 \text{ remainder}$$

This explanation is the same as the process in short division.

In this example the divisor is contained in the first figure of the dividend.

2. 7,894 by 9.

Process

$$
\begin{array}{r}
877\frac{1}{9} \\
9\overline{)7894} \\
\underline{72} \\
69 \\
\underline{63} \\
64 \\
\underline{63} \\
1
\end{array}
$$

Explanation. — 7,894 = 7 thousands + 8 hundreds + 9 tens + 4 ones.

$$
\begin{array}{r}
8 \text{ hundreds} + 7 \text{ tens} + 7 \text{ ones} \\
9\overline{)7 \text{ thousands} + {}^{7}8 \text{ hundreds} + {}^{6}9 \text{ tens} + {}^{6}4 \text{ ones}} \\
1 \text{ remainder}
\end{array}
$$

In this example the divisor is not contained in the first digit of the dividend.

305. Divide and explain the process:

1. 6)435	2. 6)371	3. 6)584	4. 6)269
5. 6)173	6. 6)6,641	7. 6)6,782	8. 6)6,923
9. 6)6,854	10. 6)6,735	11. 6)625	12. 6)634
13. 6)649	14. 6)652	15. 6)629	16. 6)7,086
17. 6)8,247	18. 6)9,488	19. 6)8,499	20. 6)9,260
21. 7)345	22. 7)538	23. 7)485	24. 7)778
25. 7)781	26. 7)7,961	27. 7)7,852	28. 7)7,643
29. 7)7,554	30. 7)7,425	31. 8)453	32. 8)358
33. 8)584	34. 8)878	35. 8)871	36. 8)9,617
37. 8)8,725	38. 8)8,173	39. 8)9,785	40. 8)9,176

41. $9\overline{)543}$ 42. $9\overline{)853}$ 43. $9\overline{)458}$ 44. $9\overline{)788}$

45. $9\overline{)178}$ 46. $9\overline{)9,176}$ 47. $9\overline{)9,278}$ 48. $9\overline{)9,173}$

49. $9\overline{)9,521}$ 50. $9\overline{)9,716}$ 51. $6\overline{)5,346}$ 52. $7\overline{)4,807}$

Division

306. Divide:

1. 87 by 10.

PROCESS

$$\begin{array}{r} 8\frac{7}{10} \\ 1\not{0}\overline{)8|7} \end{array}$$

EXPLANATION. — 87 = 8 tens + 7 ones. 10 is contained in 8 tens + 7 ones 8 times with 7 as a remainder or $8\frac{7}{10}$.

A short method of dividing a number by 10 (when the dividend does not end with a **0**) is to separate the tens from the ones by a vertical line and divide.

2. $30\overline{)127}$ 3. $40\overline{)4,326}$ 4. $50\overline{)6,534}$

5. $60\overline{)4,563}$ 6. $70\overline{)7,564}$ 7. $80\overline{)6,574}$

8. $90\overline{)9,046}$ 9. $30\overline{)2,147}$ 10. $40\overline{)8,435}$

11. $80\overline{)3,064}$ 12. $90\overline{)6,847}$ 13. $70\overline{)9,037}$

Divisors 100, 200, etc.

307. 1. Divide:

7,300 by 100.

PROCESS

$$\begin{array}{r} 73 \\ 1\not{0}\not{0}\overline{)73|\not{0}\not{0}} \end{array}$$

EXPLANATION. — 7,300 = 73 hundreds. 1 hundred is contained in 73 hundreds 73 times.

73 is the quotient.

A short method of dividing by 100 a number that

ends with two 0's is to cancel the two 0's at the right of the divisor and the dividend, and divide.

2. 3,660 by 200.

PROCESS

$$\begin{array}{r} 18\tfrac{6}{20} \\ 2\not{0}\not{0})\overline{36|60} \end{array}$$

EXPLANATION. — Write the explanation.

Explain a short method.

3. 4575 by 300.

PROCESS

$$\begin{array}{r} 15\tfrac{75}{300} \\ 3|00)\overline{45|\,75} \end{array}$$

EXPLANATION. — Write the explanation.

Canceling a cipher at the right of a number divides the number by what?

Canceling two ciphers at the right of a number divides the number by what?

308. Written Exercises

1. 100)8,400   2. 200)7,800   3. 400)97,600

4. 500)64,500   5. 600)9,600   6. 700)8,890

7. 800)89,700   8. 900)54,600   9. 600)93,310

10. 700)33,970   11. 800)828,240   12. 900)565,830

13. 700)739,256   14. 800)309,491   15. 900)267,662

16. 600)726,751   17. 500)155,671   18. 900)900,918

Oral Exercises

309. Divide by 10 and 100:

1.)500 1,300 7,500 8,000 1,200 10,000

2.)$ 1,100 $ 9,900 $ 900 $ 2,500 $3,700 $ 5,600

3.)1,000 ft. 100 ft. 2,000 lb. 2,500 lb. 5,000 mi.

Divide by 20 and 200.

4.)2,200 4,800 6,400 5,000 3,600

5.)3,200 2,600 bu. 2,800 pt. 1,200 in. 1,800 min.

310. **Review Exercises**

1. 40)80	2. 30)120	3. 30)90
4. 20)80	5. 40)120	6. 50)150
7. 30)150	8. 50)200	9. 60)240
10. 40)240	11. 90)990	12. 80)1,040
13. 90)1,170	14. 40)680	15. 700)9,100
16. 800)13,600	17. 700)1,800	18. 900)72,900

19. Divisor 400, dividend 306,000, quotient?

20. Divisor 300, dividend 405,000, quotient?

21. Divisor 700, dividend 500,600, quotient?
22. Divisor 900, dividend 873,000, quotient?
23. Divisor 600, dividend 675,341, quotient?
24. Divisor 700, dividend 3,807,560, quotient?
25. Divisor 800, dividend 4,589,670, quotient?
26. Divisor 900, dividend 7,536,420, quotient?
27. Divisor 900, dividend 7,436,420, quotient?

Review Problems

311. 1. If 20 lb. of sugar cost $1, what is the cost of 1 lb.?

2. If 30 bu. of wheat weigh 1830 lb., what is the weight of 1 bu.?

3. A farmer pays $120 for 30 sheep. How much did he pay for each sheep?

4. If a bushel of wheat sells at 97¢, what is the selling price of 70 bu.?

5. A merchant buys 90 bu. of beans for $270. What is the price per bushel?

6. If 700 bu. of clover seed cost $1400, what is the cost of 1 bu?

7. A plumber earned $200 in 50 da. How much did he earn per day?

8. One bin in a granary contains 300 bu. of grain. How many bins will be required to hold 2,400 bu.?

9. A fruit dealer has 450 bu. of apples. He packs the apples in baskets holding 20 qt. each. How many baskets will he require?

10. A plot of ground 500 ft. wide is cut up into 20 building lots of equal width. How wide is each lot?

11. A superintendent divides 1,640 pupils into 40 classes, each class containing the same number of pupils. How many pupils in each class?

12. A baker uses 30 bbl. of flour in a week. In how many weeks will he use 1,560 bbl.?

13. An express agent delivers 1,000 packages a day. Each wagon delivers 50 packages a day. How many wagons has he?

14. A family spends $420 in 40 wk. How much is spent in 1 wk. at the same rate?

15. The monthly rent receipts from an apartment house are $2,400. If there are 60 apartments, what is the average monthly rent for each apartment?

Multipliers of Two Digits

312. Multiply:

1. 23 by 13.

PROCESS

```
 23  Multiplicand
 13  Multiplier = 1 ten + 3 ones
 69 = 23 multiplied by 3 units (first partial product)
230 = 23 multiplied by 1 ten (second partial product)
299 = 23 multiplied by 1 ten, 3 ones (product)
```

EXPLANATION. — Multiply 23 by 3 ones. Write the first partial product, placing the ones under the ones and the tens under the tens. Multiply 23 by 1 ten. Write the second partial product, placing the ones under the ones, the tens under the tens and the hundreds under the hundreds. In practice the cipher is omitted. Add the partial products to obtain the product.

2. 315 by 32.

PROCESS		EXPLANATION	
315	*Multiplicand*	315 × 2 ones	= 630
32	*Multiplier*	315 × 3 tens (30)	= 9450
630		315 × 2 ones, 3 tens	= 10080
9450			
10080	*Product*		

3. $61.97 by 49.

PROCESS	EXPLANATION	
$61.97	$61.97 × 9 ones	= 557.73
49	$61.97 × 4 tens (40)	= 2478.80
557.73	$61.97 × 4 tens + 9 ones	= $3036.53
2478.80		
$3036.53		

In multiplying dollars and cents by a whole number, it is necessary to write dollars under dollars and cents under cents. For this reason the decimal points must be placed one under the other.

Written Exercise

313. Multiply

1.	2.	3.	4.	5.
78	95	103	29	98
12	13	14	15	16

6. 49 17	7. 35 18	8. 27 19	9. 123 21	10. 194 22
11. 491 23	12. 834 24	13. 305 26	14. 705 27	15. 749 34
16. 340 84	17. 650 86	18. 770 87	19. 980 78	20. 670 97

21. $93 \times 1{,}015$ 22. $86 \times 2{,}124$ 23. $95 \times 7{,}016$

24. $83 \times 7{,}108$ 25. $87 \times 6{,}789$ 26. $77 \times 6{,}543$

Note. — In multiplying it is easier to multiply the large number by the small number. For convenience arrange it in that way.

Written Problems

314. 1. A train travels 48 hr. at the rate of 35 mi. an hour. How far does it travel?

2. If a man can travel at the rate of 17 mi. an hour on a motorcycle, how far does he travel in 16 hr.?

3. A man rides 18 mi. a day for 15 da. How many miles does he ride?

4. If cherries cost 15¢ a box, 21 boxes of cherries cost $——.

5. How many ounces are there in 25 lb.?

6. A man mails 22 lb. of merchandise in ounce packages. How much postage does he pay at the rate of 1¢ an ounce?

7. How many inches in 16 ft.?

8. How many eggs in 17 doz. eggs?

9. How many sheets of paper are there in 11 quires (24 sheets = 1 quire)?

10. What is the cost of 56 gal. of gasoline at 19¢ a gal.? Show how to point off cents to make dollars.

11. A good mixture of seed for hay production to be sown by the acre is

Timothy	15 lb.
Mammoth red clover	6 lb.
Alsike clover	4 lb.

How many pounds of seed are required for 25 acres?

12. A good seed mixture for pasture land, to be sown by the acre, is

Timothy	10 lb.
Mammoth red clover	4 lb.
Alsike clover	3 lb.
White clover	2 lb.
Kentucky bluegrass	3 lb.
Tall meadow fescue	2 lb.
Orchard grass	2 lb.

How many pounds of seed are required for 40 acres?

13. A farmer found that by spraying his apple trees he increased the average income per acre from $92 to $139. What was the increase in the income for 24 acres?

14. Cattle require about 24 lb. of dry feed per day. How many pounds will 64 head of cattle require?

15. A real estate agent sold 28 lots at an average price $377.67 per lot. How much did he receive for the lots?

16. If a ton of coal costs $6.75, what will 54 tons cost?

17. A discus thrower threw the discus 134 ft. How many inches did he throw it?

18. A competitor threw the discus 133 ft. 6 in. How many inches did he throw it?

19. In an athletic contest a boy put the shot 32 ft. 4 in. What is the distance in inches?

20. A steward purchased the following goods from the grocer:

12 bu. potatoes @ 84¢ per bushel.
2 gal. molasses @ 36¢ per gallon.
56 lb. butter @ 40¢ per pound.
6 doz. eggs @ 36¢ per dozen.

What is the amount of the bill?

21. Make out a bill and receipt it.

Areas

315. 1. Draw a square figure 1 in. on each side. This figure is called a one-inch square. It has square corners. Its area is one square inch.

2. Cut from paper 2 one-inch squares.

3. Place 2 one-inch squares so as to make a figure, with square corners, twice as long as the one-inch square. Its area is 2 square inches.

4. With a rule draw a figure 3 in. long and 2 in. wide. Make the corners square corners.

5. Draw lines to show how many square inches there are in this figure.

6. Draw on the blackboard a rectangle 2 ft. wide and 4 ft. long.

7. Find the area in square feet.

8. Name objects in the room that are in the form of a rectangle.

9. Estimate the area of the surface of these objects.

10. Measure and find the area of the door. Of the window. Of the blackboard. Of this book.

Problems and Exercises

316. 1. A flower bed is 4 yd. long and 2 yd. wide. How many square yards in the surface of the bed?

2. How many strips of cloth, 1 yd. wide and 4 yd. long, will be required to cover the entire bed?

3. Can you draw a figure using 1 in. for 1 yd. to represent this flower bed?

4. Cut from a sheet of paper a rectangle 4 in. long and 2 in. wide. Divide it into square inches. What is the area?

5. Draw on the blackboard a square 1 ft. on each side. This figure is called a one-foot square. Its area is 1 square foot.

6. In the school garden each plot is 12 ft. long and 6 ft. wide. How many square feet in each plot?

7. Draw a rectangular figure, 12 ft. long and 6 ft. wide, using 1 in. to represent 1 ft.

8. Draw lines to show how many square feet there are in the figure.

9. How many square feet in top row?

10. How many square feet in each row?

11. How many rows are there?

12. What is the area of the figure?

13. Make a drawing to represent the top of a table 6 ft. long, 5 ft. wide. Use 1 in. to represent 1 ft.

14. What is the area of this rectangle?

15. Using the yardstick, measure the length and breadth of the classroom.

16. The area of a square that is 1 yd. on each side is one square yard. How many square yards in a surface 3 yd. long and 2 yd. wide? Draw diagram to illustrate.

Drill Device — Areas and Perimeters of Squares, Rectangles

317. 1. Using the diagram, draw a rectangle 5 ft. long and 3 ft. wide. Measure 5 ft. from the lower left corner to the right. From that point measure 3 ft. up on the perpendicular line. Then measure 5 ft. to the left to a point 3 ft. directly above the first corner. The enclosed space is the required rectangle.

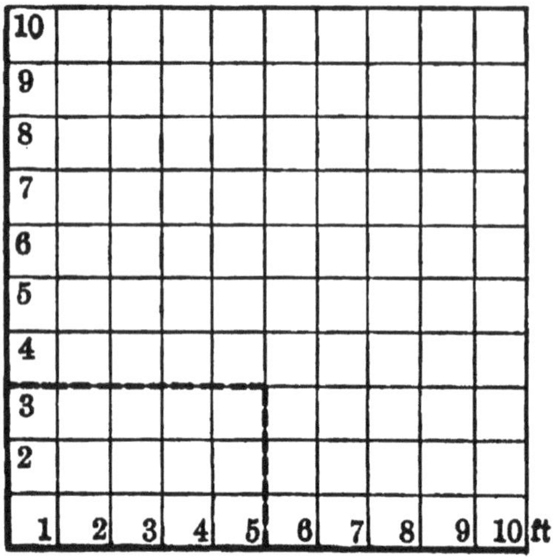

2. In this same way draw rectangles as follows:

Lengths	4	5	7	9	6	8	5	8
Widths	3	3	2	7	2	5	5	8

3. Which of these rectangles are squares?

4. Find the area in square feet of each of the rectangles.

5. Construct a similar diagram on the blackboard. Drill on rectangles and their areas.

6. Drill on finding the areas of a rectangle: When the length and width are in inches. In feet. In yards. When the dimensions are of different denominations.

7. On the blackboard draw a rectangle 4 ft. long and 2 ft. wide. How many sides has it? What is the length of each side? How many feet in the total length of the bounding lines or sides?

Process. — 2 sides, each 3 ft. long = 6 ft.
2 sides, each 2 ft. long = 4 ft.
The total length of sides or perimeter = 10 ft.

8. Find the perimeter of the following rectangles:

Lengths	5	7	9	11	6	8	10
Widths	4	3	5	6	2	7	9

9. Tell how to find the area of a rectangle.

10. Tell how to find the perimeter of a rectangle.

Oral Problems

318. 1. What is the area of a rectangle 6 in. long and 3 in. wide?

2. What is the area of a rectangle 3 yd. long and 3 yd. wide?

3. A checkerboard is 8 in. on each side. How many square inches in its surface?

4. A schoolroom is 20 ft. long and 30 ft. wide. What is its area?

5. A window is 3 ft. wide and 5 ft. high. What is its area?

6. A barn door is 12 ft. high and 9 ft. wide. Find its area.

7. How many yards of lace must be purchased in order to trim a cover that is 4 yd. long and 3 yd. wide?

8. From a piece of cloth 2 yd. long and 1 yd. wide, a piece 9 in. long and the width of the cloth was cut. What is the length of the piece that is left?

9. How many square inches in the top of a box 10 in. by 8 in.?

10. The area of the surface of a desk is 72 sq. in. One side measures 8 in. What is the length of the other side?

11. How many pieces of velvet, each $\frac{1}{2}$ yard long, can be cut from 12 yd.?

12. In the floor of a schoolroom there are 96 sq. yd. The length is 12 yd. What is the width?

13. The total length of the 4 sides of a rectangle is 36 ft. One side measures 10 ft. What is the length of one end?

14. The length of the 4 sides of a square is 40 ft. How many feet on each side?

Written Exercise

319. Find the areas of the following rectangles:

1. 5 in. by 70 in.
2. 8 in. by 90 in.
3. 7 in. by 70 in.
4. 6 ft. by 7 ft.
5. 10 ft. by 10 ft.
6. 11 ft. by 12 ft.

7. 12 ft. by 12 ft.
8. 9 yd. by 7 yd.
9. 9 yd. by 12 yd.
10. 6 in. by 80 in.

320. Find the missing dimensions in the rectangles:

	Length	Width	Area
1.	12 ft.	?	72 sq. ft.
2.	10 ft.	?	90 sq ft.
3.	?	10 ft.	100 sq. ft.
4.	?	12 ft.	144 sq. ft.
5.	18 in.	90 in.	?
6.	15 in.	70 in.	?
7.	?	7 yd.	63 sq. yd.
8.	?	9 yd.	153 sq. yd.
9.	12 yd.	?	108 sq. yd.
10.	9 yd.	8 yd.	?

321. Find the missing dimensions in the rectangles:

	Length of One Side	Length of the Other Side	Perimeter
1.	10 ft.	8 ft.	?
2.	15 ft.	?	50 ft.
3.	?	10 yd.	44 yd.
4.	20 yd.	?	70 yd.
5.	?	12 in.	52 in.

Square Measure — Exercise

322. 1. How many square ft. in 1 sq. yd.?

2. How many square inches in 1 sq. ft.?

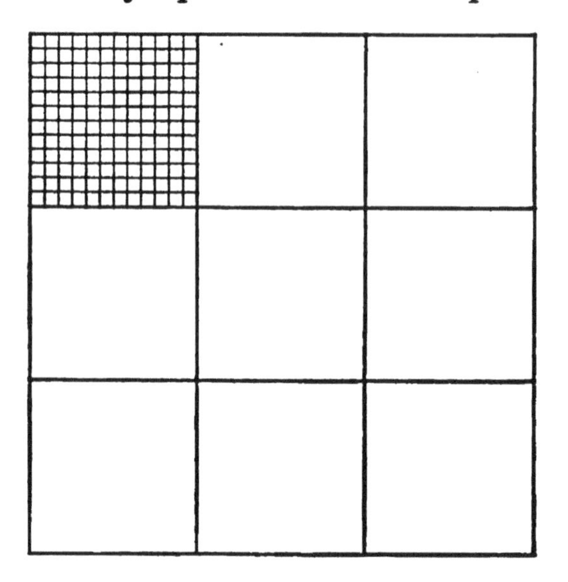

TABLE OF SQUARE MEASURE	
144 square inches (sq. in.)	= 1 square foot (sq. ft.)
9 square feet	= 1 square yard (sq. yd.)

Written Problems

323. 1. How many square inches are there in 2 sq. ft.? In 3 sq. ft.?

2. The page of a stamp album is 10 in. long and 8 in. wide. The page is divided into squares 1 in. on each side. How many of these squares are there on one page?

3. A baseball diamond is almost a square, 90 ft. on each side. What is the area in square feet?

4. If the baseball diamond is a square, 90 ft. on each side, what is the area in square yards?

5. If a baseball diamond is about 90 ft. on each side, how many yards must a player travel to make a home run?

6. An athletic field, in the form of a rectangle, is 220 yd. long and 100 yd. wide. How many square yards in the field?

7. What is the length of the perimeter of a field 220 yd. by 100 yd.?

8. The school playground is a rectangle 100 yd. long. Its area is 18,600 sq. yd. What is the width?

9. A strawberry bed is a rectangle 30 ft. long and 21 ft. wide. How many feet of chicken wire are required to fence in the bed?

10. A sign painter has an order to paint a rectangular sign board, 5 yd. long and 2 yd. wide. He paints a narrow black border. What is the length of the border? What is the area of the sign?

11. The foundation of a house is a rectangle 24 ft. by 84 ft. Find the area in square feet. In square yards.

12. The rectangular floor of a tent is 15 ft. by 18 ft. How many square feet in the floor of the tent?

13. On the floor of this tent there are 2 rectangular cots, each 3 ft. by 6 ft., a table 2 ft. square, and 4 chairs, each covering 1 sq. ft. of space. How much unoccupied space is there?

14. A football field is a rectangle 330 ft. long and 160 ft. wide. What is the area in square feet?

15. How many yards long is this field? If lines are drawn across the field at intervals of 5 yd., how many of these lines will there be?

Divisors 10, 20, 30, etc. — Review

324. Divide:

1. 471 by 20.

PROCESS

$$\begin{array}{r} 23\frac{1}{2}\!\!\!/\frac{1}{0}\!\!\!/ \\ 2\not{0})\overline{47|1} \end{array}$$

EXPLANATION. — Separate the tens and hundreds from the ones.

471 = 47 tens 1 one.

20 = 2 tens.

2 tens is contained in 47 tens 23 times and 1 ten remainder.

1 ten and 1 one remainder = 11 remainder.

2. $20\overline{)576}$	3. $30\overline{)687}$	4. $40\overline{)788}$	5. $50\overline{)891}$
6. $60\overline{)986}$	7. $70\overline{)895}$	8. $80\overline{)984}$	9. $90\overline{)999}$
10. $90\overline{)946}$	11. $80\overline{)991}$	12. $70\overline{)909}$	13. $60\overline{)899}$
14. $50\overline{)1,469}$	15. $40\overline{)2,991}$	16. $60\overline{)3,489}$	17. $70\overline{)4,899}$

Divisors of Two Figures not Ending in a Cipher

325. Divide:

1. 687 by 31.

```
PROCESS
   22 5/31
31)68|7
   62
   ×67
   ×62
     5
```

EXPLANATION. — Separate the dividend so as to show the first partial dividend. 68 is the first partial dividend.

(To find the first figure in the quotient divide 68 by 31. Try 68 ÷ 30.)

68 ÷ 31 = 2 and a remainder. Place the 2 over the right-hand figure in the partial dividend.

2 × 31 = 62; 68 − 62 = 6, the remainder.
Bring down the 7 from the dividend.
67 is the second partial dividend.
(To find the second figure in the quotient, try 67 ÷ 30.)
67 ÷ 31 = 2 and a remainder.
2 × 31 = 62; 67 − 62 = 5, the last remainder.
The quotient is 22 and the remainder is 5, or $22\frac{5}{31}$ *Ans.*

2. 43,641 by 71.

```
PROCESS
    614 47/71
71)436|41
   426
    104
     71
    331
    284
     47
```

EXPLANATION. — Separate the dividend so as to show the first partial dividend.

436 is the first partial dividend. (Try 436 ÷ 70.) 436 ÷ 71 = 6 and a remainder. Place the 6 above the right-hand figure in the partial dividend. 6 × 71 = 426; 436 − 426 = 10, the remainder. Bring down the 4, the next figure in the dividend. 140

is the second partial dividend. (Try 104 ÷ 70.) 104 ÷ 71 = 1 and a remainder. 1 × 71 = 71; 104 − 71 = 33, the remainder. Bring down the 1, the next figure in the remainder. (Try 331 ÷ 70.) 331 ÷ 71 = 4 and a remainder. 4 × 71 = 284; 331 − 284 = 47, the last remainder.

The quotient is 614 and the remainder 47, or $614\frac{47}{71}$ *Ans.*

3. $1,564.81 by 91.

PROCESS

```
         17.19 52/91
91)$ 1564.81
      91
      654
      637
       178
        91
        871
        819
         52
```

EXPLANATION. — Write the explanation. Be sure to place the decimal point in the quotient above the decimal point in the dividend.

326. **Written Exercises**

1. 31)708	2. 41)640	3. 91)980
4. 51)844	5. 31)1,708	6. 41)1,640
7. 91)1,980	8. 51)1,844	9. 61)4,879
10. 71)5,891	11. 81)7,881	12. 91)8,119

13. $21\overline{)\$5.24}$ 14. $31\overline{)\$695}$ 15. $41\overline{)\$8.78}$

16. $51\overline{)\$9.18}$ 17. $61\overline{)\$12.64}$ 18. $71\overline{)\$23.41}$

19. $81\overline{)\$36.59}$ 20. $91\overline{)\$45.79}$ 21. $51\overline{)\$109.65}$

22. How many times does 374 contain 11?

23. How many times does 2,572 contain 11?

24. Divide 8,442 into 21 equal parts.

25. Divide 7,645 into 51 equal parts.

Divisors of 2 Digits, Ones' Figure Greater than 1

327. Divide:

1. 897 by 32.

PROCESS

$$\begin{array}{r} 28\frac{1}{32} \\ 32\overline{)897} \\ \underline{64} \\ 257 \\ \underline{256} \\ 1 \end{array}$$

EXPLANATION. — $89 \div 32 = 2$ and 25 remainder. Write the 2 above the right-hand figure of the first partial dividend. Bring down the 7, the next figure in the dividend. $257 \div 32 = 8$ and 1, the last remainder.

$28\frac{1}{32}$ *Ans.*

2. 1,748 by 23.

PROCESS

$$\begin{array}{r} 76 \\ 23\overline{)1748} \\ \underline{161} \\ 138 \\ \underline{138} \end{array}$$

EXPLANATION. — Indicate the first partial dividend.

To obtain the first figure in the quotient, try $174 \div 20$.

$174 \div 20 = 8$ and a remainder. $8 \times 23 = 184$, a number larger than the first partial dividend. The first figure must be less than 8. Try 7.

$7 \times 23 = 161$; $174 - 161 = 13$, the remainder.

Bring down the next figure in the dividend. Determine in the same way the next figure in the quotient.

328. **Written Exercise**

1. $32\overline{)797}$	2. $42\overline{)869}$	3. $52\overline{)1,684}$
4. $62\overline{)1,347}$	5. $22\overline{)9,157}$	6. $72\overline{)8,095}$
7. $82\overline{)2,990}$	8. $92\overline{)9,987}$	9. $23\overline{)815}$
10. $33\overline{)697}$	11. $43\overline{)892}$	12. $53\overline{)596}$
13. $63\overline{)4,501}$	14. $93\overline{)4,645}$	15. $83\overline{)4,887}$
16. $73\overline{)5,804}$	17. $92\overline{)\$73.60}$	18. $72\overline{)\$42.48}$
19. $93\overline{)\$10.23}$	20. $73\overline{)\$83.95}$	21. $63\overline{)\$90.09}$

Written Problems

329. 1. A piece of ribbon 84 in. long is cut into pieces, each 21 in. long. How many pieces are there?

2. A carpenter cuts a piece of molding 65 in. long into pieces each 30 in. long. How many pieces are there? How long is the piece that is left over?

3. A confectioner makes 625 sticks of candy. He puts them up in jars, each holding 40 sticks. How many jars does he fill? How many sticks of candy remain?

4. A proof reader reads 20 pages of manuscript in an hour. How many hours will it take him to read 380 pages?

5. How many feet in 5,244 in.?

6. How many dozens in 4,416 eggs?

7. If a clock is wound every 21 da., how often is it wound in 357 da.?

8. How many boxes are required to hold 1,656 pencils, if each box contains 6 doz.?

9. A boat sails at the rate of 23 mi. an hour. In how many hours will it sail 2,415 mi.?

10. A boat sails at the rate of 33 mi. an hour. How many hours will it take to sail 1,617 mi.?

11. A clerk earns $22 a week. In how many weeks will he earn $418?

12. A set of 43 books costs $40.85. What is the cost of 1 book?

13. A contractor employs 93 men. The payroll for 1 wk. is $1,495.44. How much does he pay each man?

14. A dealer sells 73 bbl. of apples for $492.75. What is the cost of 1 bbl.?

Written Exercise

330. Divide:

1. 9763 by 77.

PROCESS

```
   126 61/77
77)9763
   77
   206
   154
    523
    462
     61
```

EXPLANATION. — The first partial dividend is 97. 97 ÷ 77 = 1 and 20 remainder. Bring the next digit in the dividend.

The second partial dividend is 206. Try 80 as a trial divisor for the next figure in the quotient. 206 ÷ 80 = 2.

206 ÷ 77 = 2 and 52 remainder. Bring down the last digit in the dividend. The last partial dividend is 523. Try 80 as a trial divisor for the next figure in the quotient. 523 ÷ 80 = 6.

523 ÷ 77 = 6 and 61 remainder.

$126\frac{61}{77}$ *Ans.*

2. 27,547 by 68.

PROCESS

```
    405 7/68
68)27547
   272
     347
     340
       7
```

EXPLANATION. — 275 is the first partial dividend. 275 ÷ 68 = 4 and 3 remainder.

34 is the second partial dividend. 68 is not contained in 34. Write a cipher in the quotient to indicate that the second partial dividend does not contain the divisor. Bring down the next figure in the dividend. 347 is the last partial dividend.

347 ÷ 68 = 5 and 7 remainder.

$405\frac{7}{68}$ *Ans.*

3. $158.22 by 54.

PROCESS

```
        2.93
54)$158.22
     108
      502
      486
       162
       162
```

EXPLANATION. — The first partial dividend is 158. Trial division indicates that 54 is contained in 153, 3 times. 3 × 54 = 162. 162 is greater than the first partial dividend. 158 ÷ 54 = 2 and 50 remainder. The second partial dividend is 502. 502 ÷ 54 = 9 and 16 remainder. The third partial dividend is 162. 162 ÷ 3 = 3. $2.93 *Ans.*

331. Written Exercise

1. $14\overline{)910}$	2. $15\overline{)240}$	3. $16\overline{)157}$
4. $17\overline{)187}$	5. $18\overline{)279}$	6. $23\overline{)253}$
7. $24\overline{)792}$	8. $25\overline{)960}$	9. $26\overline{)982}$
10. $27\overline{)1,607}$	11. $28\overline{)2,202}$	12. $33\overline{)990}$
13. $34\overline{)3,057}$	14. $35\overline{)1,200}$	15. $36\overline{)1,728}$
16. $37\overline{)3,885}$	17. $38\overline{)3,952}$	18. $43\overline{)4,558}$
19. $44\overline{)9,064}$	20. $45\overline{)13,815}$	21. $47\overline{)4,794}$
22. $48\overline{)2,400}$	23. $56\overline{)2,800}$	24. $57\overline{)2,850}$

Written Exercises

332. Find the quotients:

1. 96)4,535	2. 86)8,350	3. 76)31,932
4. 97)49,373	5. 87)652,863	6. 78)14,600
7. 48)$ 117.60	8. 39)$ 102.57	9. 51)$ 617.30
10. 69)$ 142.83	11. 74)$ 740.74	12. 75)$ 166.50

Written Problems

333. 1. The dividend is 1,234, the divisor 12. Find the quotient.

2. The dividend is 1,152, the quotient, 96. What is the divisor?

3. Find the quotient, when the divisor is 92 and the dividend 7,005.

4. How many feet in 276 in.? In 372 in?

5. How many pounds in 176 oz.? In 248 oz.?

6. Reduce 756 oz. to pounds and ounces.

7. In 4,572 eggs, how many dozen are there?

8. In 2,160 sheets of paper, how many quires are there (24 sheets = 1 quire)?

9. In 400 dimes how many dollars? In 800 5-cent pieces how many dollars?

10. A man earns $ 2,600 in 52 wk. What is his weekly salary?

11. A manuscript has 25,000 words. How many pages will this make allowing 300 words to a page?

12. How many times can a 40-gal. barrel be filled from a 520-gal. tank?

13. If 13 bbl. of flour weigh 2,288 lb., what is the weight of 1 bbl.?

14. A sack of coffee weighing 125 lb. sells for $18.75. What is the cost of 1 lb.?

15. A farmer paid $8,000 for 160 acres of land. He sold it at $60 an acre. What was the gain per acre? What was the total gain?

16. A real estate company bought 42 lots for $22,470. They sold them at $600 apiece. What was the gain on each lot? What was the total gain?

17. A man sold 97 crates of peaches for $232.80. On each crate he gained 40¢. What was the cost of the peaches?

To Find a Part of a Number

334. 1. Find $\frac{1}{12}$ of 276.

PROCESS

```
   23
12)276
   24
    36
    36
```

EXPLANATION. — To find $\frac{1}{12}$ of 276, divide 276 by 12.

2. $\frac{1}{15}$ of 120　3. $\frac{1}{16}$ of 180　4. $\frac{1}{12}$ of 360
5. $\frac{1}{10}$ of 960　6. $\frac{1}{11}$ of 4,323　7. $\frac{1}{13}$ of 3,900
8. $\frac{1}{14}$ of 1,960　9. $\frac{1}{15}$ of 3,045　10. $\frac{1}{16}$ of 4,865
11. $\frac{1}{24}$ of 1,696　12. $\frac{1}{36}$ of 4,800　13. $\frac{1}{75}$ of 5,700
14. $\frac{1}{31}$ of 32,886　15. $\frac{1}{96}$ of 48,960　16. $\frac{1}{50}$ of 7,500

Divisors of 2 Figures, Ones' Figure 9

335. Divide:

1. 2,790 by 29.

PROCESS

```
    96 6/29
29)2790
   261
    180
    174
      6
```

EXPLANATION. — 279 is the first partial dividend. Use 30 for trial divisor. 279 ÷ 30 = 9. 9 × 29 = 261; 279 − 261 = 18 remainder. The partial dividend is 180. 180 ÷ 30 = 6. 6 × 29 = 174; 180 − 174 = 6 remainder.

$96\frac{6}{29}$ *Ans.*

When the ones' figure is 9, as 29, 39, etc., it is generally easier to use 30, 40, etc., as a trial divisor.

2. 2,499 by 49.

PROCESS

```
    51
49)2499
   245
     49
     49
```

EXPLANATION. — 249 is the first partial dividend. Use 50 for trial divisor. 249 ÷ 50 = 4. 4 × 49 = 196; 249 − 196 = 55 remainder. When the remainder is greater than the divisor, the divisor will be contained in the dividend a greater number of times. In this case the trial divisor, 50, gives an incorrect quotient. 5 is the first figure in the quotient.

336. **Written Exercises**

1. $39\overline{)498}$	2. $49\overline{)1,251}$	3. $59\overline{)2,039}$	4. $69\overline{)6,506}$
5. $79\overline{)1,848}$	6. $89\overline{)4,058}$	7. $99\overline{)7,612}$	8. $19\overline{)1,742}$
9. $29\overline{)8,787}$	10. $49\overline{)9,898}$	11. $79\overline{)7,979}$	12. $99\overline{)9,999}$

337. Divide:

1. 1,476 by 12.

LONG DIVISION PROCESS

$$\begin{array}{r} 123 \\ 12\overline{)1476} \\ \underline{12} \\ 27 \\ \underline{24} \\ 36 \\ \underline{36} \end{array}$$

SHORT DIVISION PROCESS

$$\begin{array}{r} 1\ 2\ 3 \\ 12\overline{)14^{2}7^{3}6} \end{array}$$

EXPLANATION

$$\begin{array}{r} 1\text{ h} + 2\text{ t} + 3\text{ o} \\ 12\overline{)1476 = 14\text{ h} + {}^{2}7\text{ t} + {}^{3}6\text{ o}} \end{array}$$

Written Exercise

338. Divide by short division:

1. $11\overline{)946}$	2. $11\overline{)5,467}$	3. $11\overline{)7,645}$	4. $11\overline{)902}$
5. $11\overline{)8,470}$	6. $12\overline{)867}$	7. $12\overline{)908}$	8. $12\overline{)875}$
9. $12\overline{)1,875}$	10. $12\overline{)648}$	11. $11\overline{)1,080}$	12. $12\overline{)1,080}$

Find $\frac{1}{11}$ of the following numbers:

13. 253 **14.** 391 **15.** 594 **16.** 1,012 **17.** 7,337

Find $\frac{1}{12}$ of the following numbers:

18. 1,440 **19.** 1,728 **20.** 5,280 **21.** 2,871 **22.** 4,410

Find the answer:

23. $\frac{1}{12}$ of $ 374.40 **24.** $\frac{1}{11}$ of $ 44.58 **25.** $\frac{1}{10}$ of $ 1,000

Analysis — Oral

339. 1. If 6 oranges cost 18 ¢, what will 1 orange cost?

2. If 5 pencils cost 25 ¢, what will 1 pencil cost? What will 7 pencils cost?

3. If 2 boxes of candy cost 50 ¢, what will 1 box cost? What will 3 boxes cost?

4. If 2 pears cost 6 ¢, what will 3 pears cost?

5. If a boy on a bicycle travels 20 miles in 2 hours, how far will he travel in 3 hours? In 7 hours?

6. What does a man earn in 5 days, if in six days he earns $ 18?

Analysis — Written

340. 1. If 20 lb. of nails cost 80 ¢, what will 1 lb. cost? What will 17 lb. cost?

2. What is the cost of 1 bu. of apples, if 11 bu. cost $ 18.70?

3. A pound contains 16 ounces; how many pounds in 368 ounces?

4. If 8 yd. of cloth cost $ 22.40, what does 1 yd. cost? What do 15 yd. cost?

5. If 135 bushels of wheat are raised on 5 acres of land, how many bushels are raised per acre? How many bushels are raised on 4 acres?

6. If 11 tons of lignite coal are sold at $ 38.50, what is the price of 1 ton? 6 tons?

7. If 2 gallons of ice cream are served to 48 people, then 3 gallons of ice cream can be served to —— people.

8. If a man earns $ 525 in 7 months, how much does he earn in 1 month? How much in 4 months?

9. A farmer planted 360 hills of potatoes in 6 rows. How many hills did he plant in 1 row? How many in 7 rows?

10. If a farmer sells 8 bu. of potatoes for $ 6.40, what is the price per bushel? What is the price of 5 bu.?

11. Find the cost of 3 doz. eggs when 4 doz. cost $ 1.80.

12. If 2 doz. spools of cotton sell for $ 1.10, what do 3 doz. sell for?

13. What is the weight of 3 bu. of potatoes, if 4 bu. of potatoes weigh 240 lb.?

Review — Oral

341. 1. Repeat the multiplication table of 8's, of 7's, of 9's.

2. Answer at sight:

6×8 $\quad 7 \times 9$ $\quad 8 \times 7$ $\quad 6 \times 7$ $\quad 5 \times 8$

9×9 $\quad 9 \times 8$ $\quad 6 \times 9$ $\quad 7 \times 7$ $\quad 8 \times 8$

9)63 $\quad$ 8)56 $\quad$ 7)42 $\quad$ 8)72 $\quad$ 11)121

12)72 $\quad$ 12)84 $\quad$ 11)110 $\quad$ 9)81 $\quad$ 8)64

3. State the areas of the following rectangles:

Length in feet: 11, 12, 9, 13, 12, 3, 4.
Width in feet: 10, 11, 11, 10, 12, 12, 22.

4. How many inches in 1 ft.? In 3 ft.? In 7 ft.?

5. How many feet in 12 in.? In 24 in.? In 48 in.?

6. How many days in 2 weeks? In 6 weeks? In 7 weeks?

7. How many weeks in 7 days? In 21 days? In 35 days?

342. **Written Review**

1. 12)150 $\quad$ 12)100 $\quad$ 12)245 $\quad$ 12)370 $\quad$ 12)850

2. 31)325 $\quad$ 31)706 $\quad$ 13)325 $\quad$ 25)750 $\quad$ 30)6,090

3. 19×74 $\quad 91 \times 74$ $\quad 23 \times 86$ $\quad 98 \times 15$ $\quad 103 \times 41$

4. $5 \times \$18.50$ $\quad 11 \times \$12.25$ $\quad 12 \times \$5.85$

PROPERTIES OF WHOLE NUMBERS

Factors

343. 1. What two numbers are there whose product is 10, 14, 15, 25, 33, 35, 39?

2. What three numbers are there whose product is 12, 18, 24, 27, 32, 44, 56?

3. What two numbers multiplied together produce 45, 56, 64, 72, 96?

4. What three numbers multiplied together produce 45, 56, 64, 72, 81, 96?

5. What four numbers multiplied together produce 64, 72, 81, 96?

The whole numbers which when multiplied together produce a number are called the **factors** of the number.

Oral Exercise

344. Name the factors and the products.

1.	$3 \times 3 = 9$	2.	$2 \times 7 = 14$
3.	$3 \times 5 = 15$	4.	$2 \times 9 = 18$
5.	$3 \times 7 = 21$	6.	$2 \times 11 = 22$
7.	$2 \times 2 \times 2 = 8$	8.	$2 \times 3 \times 5 = 30$
9.	$2 \times 3 \times 7 = 42$	10.	$4 \times 6 \times 3 = 72$
11.	$4 \times 8 \times 2 = 64$	12.	$2 \times 2 \times 2 \times 2 = 16$

345. Find all the missing factors:

1. $4 \times ? = 12$	2. $? \times 6 = 48$
3. $7 \times ? = 56$	4. $? \times 9 = 98$
5. $8 \times ? = 64$	6. $? \times 8 = 72$
7. $? \times ? \times 7 = 28$	8. $? \times 9 = 45$
9. $8 \times ? \times 5 = 120$	10. $? \times ? \times 2 \times ? = 32$

The factors of a number divide the number exactly, that is, without remainder. They are sometimes called **exact divisors**.

A number that has 2 as a factor is called an **even number**.

Even numbers have 2, 4, 6, 8, or 0 in ones' place. 12, 14, 16, 18, 20, etc., are even numbers.

A number that does not contain 2 as a factor is called an **odd number**.

Odd numbers have 1, 3, 5, 7, 9 in ones' place. 11, 13, 15, 17, 19, etc., are odd numbers.

Divisibility of Numbers

346. 1. When is a number exactly divisible by 2?

2. When is a number exactly divisible by 3?

3. When is a number exactly divisible by 5?

4. When is a number exactly divisible by 10?

5. If the ones' figure of a number is 2, 4, 6, 8, or 0, it is exactly divisible by 2, *e.g.* 82, 94, 96, 98, 100.

If the sum of its figures is divisible by 3, the number is exactly divisible by 3, *e.g.* 312, 129, 108.

If the ones' figure is 0 or 5, the number is exactly divisible by 5, *e.g.* 25, 30, 35, 40.

If the ones' figure is 0, the number is exactly divisible by 10, *e.g.* 10, 20, 100.

Select the numbers that are exactly divisible by 2, 3, 5 and 10.

48	50	35	47	72
100	125	165	49	25
148	150	162	201	990
765	270	750	462	6400

Problems

347. 1. Can 16 oranges be divided equally among three persons without cutting any of the oranges?

2. Can $43 be spent exactly in the purchase of sheep at $5 each?

3. A father has 8 one-dollar bills. Can he divide this sum equally among 5 children, without making change?

4. A man has $50. He wants to spend all of it in buying chairs at $5 each. Can he do it? Can he do it if he pays $6 each?

5. Can a man spend all of $166 in buying hats at $3 each?

6. John's stride is 3 ft. Can he pace 213 ft. and exactly reach the end?

7. Can you make up 312 pencils into packages of half a dozen each, and have no pencils left?

8. Is the length of a yardstick contained an exact number of times in a fence 1035 ft.?

PART THREE

CANCELATION

Oral Exercise

348. 1. How much is 24 divided by 6? 8 divided by 2?

24 divided by 6 is sometimes written $\frac{24}{6}$.

In the same way, 8 divided by 2 is sometimes written $\frac{8}{2}$.

2. Compare $\frac{24}{6}$ with $\frac{8}{2}$ and their quotients.

In $\frac{24}{6} = 4$ and $\frac{8}{2} = 4$ the dividend, 24, is 3 times the dividend 8 and the divisor, 6, is 3 times the divisor 2. The quotient is the same in both divisions. This shows that dividing both dividend and divisor of $\frac{24}{6}$ by their common factor 3 does not change the quotient. That is,

$$\frac{\overset{8}{\not{2}\not{4}}}{\underset{2}{\not{6}}} = \frac{8}{2} = 4.$$

3. Divide dividend and divisor by their common factors: $\frac{70}{60}$, $\frac{75}{35}$, $\frac{72}{46}$, $\frac{72}{39}$, $\frac{55}{50}$.

Cancelation

349. 1. Divide 12×15 by 8×10.

PROCESS BY CANCELATION

$$\frac{12 \times 15}{8 \times 10} = \frac{\overset{3}{\cancel{12}} \times 15}{\underset{2}{\cancel{8}} \times \underset{2}{\cancel{10}}} \quad \begin{matrix}\text{(Dividend)} \\ \text{(Divisor)}\end{matrix}$$

(3 written above 15 in the dividend.)

$$\frac{\overset{3}{\cancel{12}} \times \overset{3}{\cancel{15}}}{\underset{2}{\cancel{8}} \times \underset{2}{\cancel{10}}} = \frac{3 \times 3}{2 \times 2} = \frac{9}{4} = 2\tfrac{1}{4}$$

EXPLANATION. — Take the common factor 4 out of 12 in the dividend and 8 in the divisor, leaving the factors 3 and 2.

Take the common factor 5 out of 15 in the dividend and 10 in the divisor, leaving the factors 3 and 2.

Find the product of 3×3, the remaining factors in the dividend, and 2×2, the remaining factors in the divisor, and divide.

2. Divide, using cancelation, $\frac{12 \times 15}{10 \times 18}$.

PROCESS

$$\frac{12 \times 15}{10 \times 18} = \frac{\overset{2}{\cancel{12}} \times \overset{3}{\cancel{15}}}{\underset{2}{\cancel{10}} \times \underset{3}{\cancel{18}}} = \frac{6}{6} = 1$$

EXPLANATION. — Reject the common factors 6 and 5 from both the dividend and the divisor.

Divide the product of the factors in the dividend by the products of the factors in the divisor.

3. Divide, using cancelation, $\frac{12 \times 15 \times 42}{8 \times 9 \times 24}$.

PROCESS

$$\frac{12 \times 15 \times 42}{8 \times 9 \times 24} = \frac{\overset{3}{\cancel{12}} \times \overset{5}{\cancel{15}} \times \overset{14}{\cancel{42}}}{\underset{2}{\cancel{8}} \times \underset{3}{\cancel{9}} \times \underset{8}{\cancel{24}}}$$

$$= \frac{\overset{\overset{}{\cancel{3}}}{\cancel{12}} \times \overset{5}{\cancel{15}} \times \overset{\overset{7}{\cancel{14}}}{\cancel{42}}}{\underset{\cancel{2}}{\cancel{8}} \times \underset{\cancel{3}}{\cancel{9}} \times \underset{8}{\cancel{24}}} = \frac{35}{8} = 4\tfrac{3}{8}$$

EXPLANATION. — Reject the common factors, 4, 3 and 3, from both the dividend and the divisor.

Reject the common factors, 3 and 2, from the new dividend and divisor. Divide the product of the remaining factors of the dividend by the product of the remaining factors of the divisor.

PRINCIPLE TO BE REMEMBERED

Dividing both dividend and divisor by the same factor does not change the value of the quotient.

Written Exercise

350. Divide, using cancelation:

1. $24 \times 49 \times 18$ by $12 \times 21 \times 36$
2. $25 \times 35 \times 56$ by $15 \times 28 \times 49$
3. $32 \times 108 \times 100$ by $64 \times 36 \times 25$
4. $39 \times 28 \times 72$ by $35 \times 52 \times 24$
5. $16 \times 40 \times 24$ by $20 \times 8 \times 48$

6. $\dfrac{30 \times 32 \times 36 \times 40}{50 \times 16 \times 20 \times 18}$

7. $\dfrac{35 \times 120 \times 72}{50 \times 63 \times 6}$

8. $\dfrac{625 \times 49 \times 81}{75 \times 21 \times 14}$

9. $\dfrac{144 \times 99 \times 10}{12 \times 88 \times 5}$

10. $\dfrac{49 \times 64 \times 70}{140 \times 16 \times 2}$

11. $\dfrac{2500}{4000}$

12. $\dfrac{1600}{3600}$

COMMON FRACTIONS

351. 1. Into how many *equal* parts is this rectangle divided?

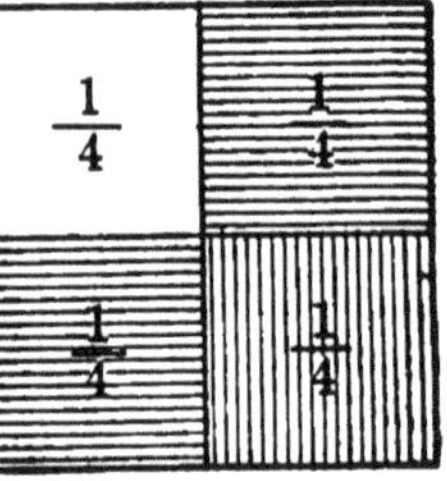

How many parts are shaded?

$\frac{3}{4}$ means that the whole rectangle is divided into —— *equal* parts, and that —— of these parts are shaded.

2. Into how many equal parts is this rectangle divided?

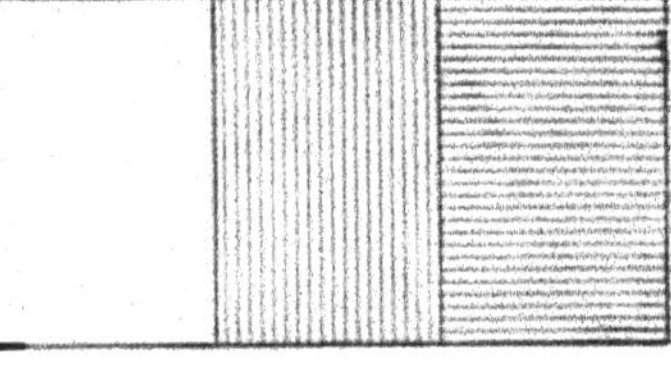

How many of these parts are shaded?

What fraction tells the total number of equal parts and the number of parts shaded?

What fraction tells the total number of equal parts and the number of parts *not* shaded?

A fraction is made up of two numbers, called *terms*.

The term below the line is called the *denominator.*

It shows into how many equal parts a unit is divided.

The term above the line is called the *numerator.*

It shows how many of the equal parts are taken.

3. Which is the denominator in $\frac{1}{2}$, $\frac{5}{6}$, $\frac{1}{6}$? Which is the numerator?

4. Make a drawing and show from it the meaning of $\frac{2}{5}$.

5. Which is larger, $\frac{5}{6}$ or $\frac{4}{6}$? Why?

Written Exercise

352. Write in figures:

1. Three fourths.
2. Seven twelfths.
3. Nine tenths.
4. Three fifths.
5. One eighth.
6. Seven eighths.
7. Three eighths.
8. Five sixths.
9. Five eighths.

Written Exercise

353. Write in words:

1. $\frac{1}{4}$ 2. $\frac{5}{12}$ 3. $\frac{7}{10}$ 4. $\frac{2}{4}$ 5. $\frac{1}{6}$

6. $\frac{4}{6}$ 7. $\frac{1}{8}$ 8. $\frac{7}{12}$ 9. $\frac{11}{12}$ 10. $\frac{7}{16}$

Halves, Fourths, and Eighths — Comparison

354. Show by diagram:

1. $\frac{1}{2}+\frac{1}{2}=1$
2. $\frac{2}{2}=1$
3. $\frac{2}{4}=\frac{1}{2}$
4. $\frac{3}{4}=\frac{1}{2}+\frac{1}{4}$
5. $\frac{1}{4}+\frac{1}{4}+\frac{1}{4}+\frac{1}{4}=1$
6. $\frac{4}{4}=1$

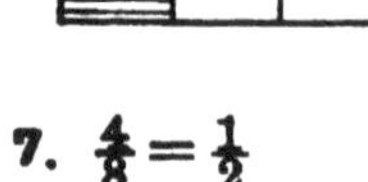

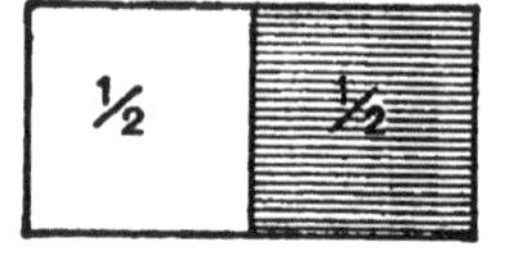

7. $\frac{4}{8}=\frac{1}{2}$
8. $\frac{3}{8}=\frac{1}{4}+\frac{1}{8}$
9. $\frac{5}{8}=\frac{1}{2}+\frac{1}{8}$
10. $\frac{6}{8}=\frac{1}{2}+\frac{1}{4}$
11. $\frac{6}{8}=\frac{3}{4}$
12. $\frac{7}{8}=\frac{1}{2}+\frac{3}{8}$
13. $1-\frac{1}{2}=\frac{1}{2}$
14. $\frac{2}{2}-\frac{1}{2}=\frac{1}{2}$
15. $\frac{2}{4}-\frac{1}{4}=\frac{1}{4}$
16. $\frac{3}{4}-\frac{1}{4}=\frac{1}{2}$
17. $\frac{3}{4}-\frac{1}{2}=\frac{1}{4}$
18. $\frac{8}{8}=1$
19. $1-\frac{7}{8}=\frac{1}{8}$
20. $1-\frac{5}{8}=\frac{3}{8}$
21. $\frac{5}{8}-\frac{1}{8}=\frac{4}{8}$
22. $\frac{6}{8}-\frac{1}{2}=\frac{1}{4}$
23. $\frac{6}{8}-\frac{1}{4}=\frac{1}{2}$
24. $\frac{7}{8}-\frac{1}{2}=\frac{3}{8}$

Oral Problems

355. 1. How much is $\frac{1}{2}$ ft. + $\frac{1}{2}$ ft.?

2. How much is $\$\frac{1}{2} + \$\frac{1}{2}$?

3. How much is $\$1 - \$\frac{1}{2}$?

4. How much is 1 bu. − $\frac{1}{2}$ bu.?

5. What is the sum of $\frac{1}{2}$ bu. and $\frac{1}{2}$ bu.?

6. What is the difference between 1 mi. and $1\frac{1}{2}$ mi.?

7. Add $\frac{1}{2}$ qt. and $\frac{1}{2}$ qt.

8. Subtract $\frac{1}{2}$ gal. from 1 gal.

9. How much is $\frac{1}{2}$ of an apple and $\frac{1}{4}$ of an apple?

10. What is the difference between one half of a dollar and one fourth of a dollar?

11. Add one fourth of a bushel and one fourth of a bushel.

12. Subtract one fourth of a pound from one pound.

Exercise

356. Copy and complete:

1. $1 - \frac{1}{2} = ?$
2. $1 - \frac{1}{4} = \frac{}{4}$
3. $1 - \frac{2}{4} = \frac{}{4} = \frac{}{2}$
4. $1 - \frac{3}{4} = \frac{}{4}$
5. $\frac{1}{2} + \frac{1}{2} = ?$
6. $\frac{1}{4} + \frac{1}{4} = \frac{}{4} = \frac{}{2}$
7. $\frac{1}{8} + \frac{1}{8} + \frac{1}{8} = \frac{}{8}$
8. $1 - \frac{1}{8} = \frac{}{8}$
9. $1 - \frac{7}{8} = \frac{}{8}$
10. $\frac{1}{8} + \frac{3}{8} = \frac{}{8}$
11. $\frac{1}{8} + \frac{3}{8} = \frac{}{4}$
12. $\frac{1}{8} + \frac{3}{8} = \frac{}{2}$
13. $\frac{3}{4} + \frac{1}{4} = ?$
14. $1 - \frac{2}{4} = ?$
15. $\frac{1}{4} + \frac{1}{4} = ?$

16. $1-\frac{3}{8}=?$ 17. $\frac{1}{4}-\frac{1}{8}=?$ 18. $\frac{3}{8}+\frac{3}{8}=?$

19. $\frac{5}{8}+\frac{3}{8}=?$ 20. $\frac{1}{2}+\frac{1}{4}+\frac{1}{4}=?$ 21. $\frac{3}{4}-\frac{1}{2}=?$

It will be found advantageous in studying these relations to use objects, paper circles, paper rectangles, etc., that may be marked, folded, or cut by children.

Halves, Thirds, Sixths — Comparison

357. Show by diagram:

1. $\frac{1}{2}=\frac{3}{6}$ 2. $\frac{3}{3}=1$ 3. $\frac{1}{2}+\frac{1}{3}=\frac{5}{6}$

4. $\frac{1}{3}=\frac{2}{6}$ 5. $\frac{6}{6}=1$ 6. $\frac{1}{3}-\frac{1}{6}=\frac{1}{6}$

7. $\frac{2}{3}=\frac{4}{6}$ 8. $\frac{1}{3}+\frac{1}{6}=\frac{1}{2}$ 9. $\frac{1}{2}-\frac{1}{3}=\frac{1}{6}$

10. $\frac{1}{6}+\frac{1}{6}+\frac{1}{6}+\frac{1}{6}+\frac{1}{6}+\frac{1}{6}=1$

Written Exercise

358. Copy and complete:

1. $\frac{1}{2}=\frac{}{6}$ 2. $\frac{1}{3}=\frac{}{6}$ 3. $\frac{2}{2}=\frac{}{3}$

4. $\frac{}{6}=1$ 5. $\frac{1}{2}=\frac{}{3}+\frac{1}{6}$ 6. $\frac{1}{2}=\frac{1}{3}+\frac{}{6}$

7. $\frac{}{2}=\frac{1}{3}+\frac{1}{6}$ 8. $\frac{2}{3}=\frac{}{6}$ 9. $\frac{1}{2}-\frac{1}{3}=\frac{}{6}$

10. $\frac{}{2}-\frac{1}{3}=\frac{1}{6}$ 11. $\frac{1}{2}+\frac{1}{6}=\frac{}{6}$ 12. $\frac{1}{3}+\frac{}{2}=\frac{5}{6}$

Oral Problems

359. 1. If 1 lb. of sugar costs 6¢, what is the cost of one half pound?

2. If 1 bu. of grain weighs 60 lb., what is the weight of one half bushel?

3. A boy received on his birthday one half dollar from his brother and one quarter from his sister. How much money did he receive?

4. A cake was divided equally among 8 people. What part of the cake did each receive?

5. From 1 yd. of ribbon a clerk cut $\frac{1}{3}$ of a yard. How much ribbon was left?

6. A dressmaker had 1 yd. of velvet. She used $\frac{1}{3}$ of a yard to make a collar and $\frac{1}{6}$ of a yard for trimmings. How much velvet had she left?

7. In a class there are 36 pupils. One sixth of them are 10 yr. old. How many pupils are 10 yr. old?

8. A baker had 18 doz. biscuits. He sold $\frac{2}{9}$ of the number at one store, and $\frac{1}{9}$ at another store. How many dozen did he sell?

9. A woman bought 1 doz. oranges. She used $\frac{2}{3}$ of them on Monday and $\frac{1}{6}$ of them on Tuesday. How many did she have left?

10. A boy had 12 marbles. He lost $\frac{5}{6}$ of the number in a game. How many marbles has he left?

Fractional Parts of Whole Numbers

360. 1. How much is $\frac{1}{4}$ of 24?

PROCESS

$$4\overline{)24}\quad 6$$

EXPLANATION. — To find one fourth of a number, divide the number by 4.

2. How much is $\frac{3}{4}$ of 24?

PROCESS

$$4\overline{)24}\ \text{quotient } 6$$

$3 \times 6 = 18$ *Ans.*

EXPLANATION. — To find three fourths of a number, divide the number by 4 and multiply the quotient by 3.

3. How much is $\frac{5}{6}$ of 1,728?

PROCESS

$$6\overline{)1728}\ \text{quotient } 288$$

$5 \times 288 = 1,440$ *Ans.*

Write the explanation.

Exercise

361. Find the answer. Use pencil only when necessary.

1. $\frac{1}{4}$ of 28	2. $\frac{3}{4}$ of 28	3. $\frac{1}{2}$ of 20
4. $\frac{1}{4}$ of 20	5. $\frac{3}{4}$ of 20	6. $\frac{3}{4}$ of 16
7. $\frac{2}{4}$ of 16	8. $\frac{1}{2}$ of 16	9. $\frac{1}{8}$ of 32
10. $\frac{3}{8}$ of 32	11. $\frac{5}{8}$ of 32	12. $\frac{7}{8}$ of 32
13. $\frac{1}{4}$ of $12	14. $\frac{3}{4}$ of $12	15. $\frac{1}{8}$ of $24
16. $\frac{3}{8}$ of $24	17. $\frac{5}{8}$ of $24	18. $\frac{7}{8}$ of $24
19. $\frac{7}{8}$ of 40	20. $\frac{7}{8}$ of 48	21. $\frac{7}{8}$ of 56
22. $\frac{3}{8}$ of 64	23. $\frac{5}{8}$ of 72	24. $\frac{7}{8}$ of 80
25. $\frac{1}{3}$ of 18	26. $\frac{2}{3}$ of 18	27. $\frac{1}{6}$ of 18
28. $\frac{5}{6}$ of 18	29. $\frac{1}{3}$ of 24	30. $\frac{2}{3}$ of 24
31. $\frac{1}{6}$ of 24	32. $\frac{2}{6}$ of 24	33. $\frac{3}{6}$ of 24

34. $\frac{4}{6}$ of 24 35. $\frac{5}{6}$ of 24 36. $\frac{1}{9}$ of 36

37. $\frac{2}{9}$ of 36 38. $\frac{3}{9}$ of 36 39. $\frac{1}{3}$ of 36

40. $\frac{4}{9}$ of 36 41. $\frac{5}{9}$ of 36 42. $\frac{6}{9}$ of 36

Oral Exercise

362. Read:

1. $5\frac{3}{4}$ (it is read five and three fourths).

2. $18\frac{7}{8}$ 3. $66\frac{2}{3}$ 4. $33\frac{1}{3}$ 5. $16\frac{1}{2}$

6. $12\frac{1}{2}$ 7. $45\frac{5}{9}$ 8. $100\frac{1}{3}$ 9. $250\frac{9}{10}$

Numbers like $5\frac{3}{4}$ are called **mixed numbers.**

Written Exercise

363. Write in figures:

1. One and one half. 2. Two and one half.
3. Three and one half. 4. Fourteen and one half.
5. Nineteen and one half.
6. Ten and one fourth.
7. Twelve and three fourths.
8. Four and three eighths.
9. Two and seven eighths.
10. Two halves.
11. Four fourths.
12. Eight eighths.
13. One hundred and two thirds.
14. Three hundred fifty-six and three fifths.
15. One thousand and five ninths.

Mixed Numbers — Addition

364. Add:

1. $3\frac{1}{2}$
   $\underline{4\frac{1}{4}}$

Process

Arrange the work as follows:

$$\begin{array}{r} 3\frac{1}{2} = 3\frac{2}{4} \\ \underline{4\frac{1}{4} = 4\frac{1}{4}} \\ 7\frac{3}{4} \end{array}$$

Explanation

$\frac{1}{2} = \frac{2}{4}$

$3\frac{1}{2} = 3\frac{2}{4}$

$3\frac{2}{4} + 4\frac{1}{4} = 7\frac{3}{4}$

2. $15\frac{2}{3}$
   $\underline{10\frac{1}{3}}$

Process

$15\frac{2}{3}$

$\underline{10\frac{1}{3}}$

$25\frac{3}{3} = 25 + 1 = 26$ *Ans.*

Explanation

$15\frac{2}{3} + 10\frac{1}{3} = 25\frac{3}{3}$

$\frac{3}{3} = 1$

$25\frac{3}{3} = 25 + 1 = 26$

Mixed Numbers — Addition

365. Add:

1.	2.	3.	4.	5.
$5\frac{2}{3}$	$3\frac{1}{4}$	$2\frac{5}{6}$	$2\frac{5}{8}$	$9\frac{1}{4}$
$3\frac{1}{3}$	$4\frac{3}{4}$	$6\frac{1}{6}$	$2\frac{3}{8}$	$1\frac{3}{8}$

6.	7.	8.	9.	10.
$9\frac{1}{2}$	$3\frac{1}{2}$	$5\frac{1}{3}$	$10\frac{1}{2}$	$6\frac{1}{2}$
$10\frac{1}{4}$	$4\frac{2}{3}$	$3\frac{1}{6}$	$8\frac{1}{6}$	$4\frac{1}{6}$

11.	12.	13.	14.	15.
$12\frac{1}{4}$	$16\frac{2}{3}$	$12\frac{1}{8}$	$15\frac{2}{3}$	$10\frac{1}{2}$
$8\frac{1}{2}$	$16\frac{2}{3}$	$12\frac{5}{8}$	$12\frac{1}{6}$	$9\frac{1}{2}$

Mixed Numbers — Subtraction

366. Subtract:

1. $7\frac{1}{2}$
   $\underline{2\frac{1}{4}}$

PROCESS

Arrange the work as follows:

$$\begin{array}{r} 7\frac{1}{2} = 7\frac{2}{4} \\ \underline{2\frac{1}{4} = 2\frac{1}{4}} \\ 5\frac{1}{4} \end{array}$$

EXPLANATION

$$\frac{1}{2} = \frac{2}{4}$$
$$7\frac{1}{2} = 7\frac{2}{4}$$
$$7\frac{2}{4} - 2\frac{1}{4} = 5\frac{1}{4}$$

Written Exercise

367. Subtract:

1.	2.	3.	4.	5.
$5\frac{1}{2}$	$3\frac{1}{2}$	$7\frac{3}{4}$	$7\frac{5}{8}$	$6\frac{5}{8}$
$\underline{4}$	$\underline{1\frac{1}{2}}$	$\underline{5\frac{1}{4}}$	$\underline{3\frac{3}{8}}$	$\underline{4\frac{1}{4}}$

6.	7.	8.	9.	10.
$5\frac{1}{2}$	$5\frac{1}{2}$	$8\frac{5}{8}$	$4\frac{7}{8}$	$6\frac{2}{3}$
$\underline{3\frac{1}{3}}$	$\underline{4\frac{1}{6}}$	$\underline{3\frac{1}{4}}$	$\underline{4\frac{3}{4}}$	$\underline{3\frac{1}{2}}$

Written Problems

368. 1. A grocer sold $13\frac{1}{2}$ lb. of cheese to one customer and $15\frac{1}{4}$ lb. to another. How many pounds did he sell to both?

2. A man spent \$35 for a suit of clothes, \$$4\frac{1}{2}$ for a pair of shoes, \$$1\frac{1}{4}$ for a tie. How much did he spend in all?

3. A clerk who earned \$25 a week spent \$$8\frac{1}{2}$ for board and \$$6\frac{1}{4}$ for other expenses. How much did he save each week?

4. A wagon loaded with hay weighs 2975 lb. The wagon when unloaded weighs $1187\frac{3}{8}$ lb. What is the weight of the hay?

5. A boy walks $\frac{5}{6}$ of a mile to school and $\frac{1}{6}$ of a mile on an errand. How far did he walk?

6. From a piece of lace $3\frac{5}{6}$ yd. long a piece $2\frac{1}{3}$ yd. long was cut. How many yards were left?

7. Two boards are $4\frac{3}{4}$ in. and $5\frac{1}{2}$ in. wide. What is the total width?

8. A class spent $\frac{2}{3}$ of an hour studying arithmetic and $\frac{1}{4}$ hr in manual training. How much time did the two periods occupy?

9. A tailor used $2\frac{3}{8}$ yd. of cloth to make a coat and $\frac{1}{2}$ yd. to make the vest. How many yards did it take to make the coat and vest?

10. If a man works $8\frac{2}{3}$ hr. a day and sleeps $8\frac{1}{6}$ hr. a day, how many hours are left for other uses?

11. A dressmaker charged \$ $12\frac{1}{2}$ for cloth, \$ $2\frac{1}{4}$ for lining and \$ $10\frac{1}{4}$ for making a dress. What was the entire cost of the dress?

12. A mechanic's earnings for 3 da. were as follows: Monday \$ $3\frac{1}{2}$, Tuesday \$ 3, Wednesday \$ $4\frac{1}{4}$. How much did he earn in the 3 da.?

13. A boy earned \$ $\frac{3}{4}$ by doing errands, \$ $\frac{1}{4}$ by cleaning the yard, \$ 1 by cleaning windows. What were his total earnings?

14. A dealer bought $8\frac{1}{3}$ doz. eggs from one farmer, $10\frac{1}{6}$ doz. from another farmer, and $6\frac{1}{3}$ doz. from a third farmer. How many dozen eggs did he buy?

Mixed Numbers — Multiplication

369. Multiply:

1. 14 by $2\frac{1}{2}$.

PROCESS

$$\begin{array}{rl} 14 & \\ \times\, 2\frac{1}{2} & \\ \hline 7 & \text{first partial product} \\ \underline{28} & \text{second partial product} \\ 35 & \text{product} \end{array}$$

EXPLANATION

$$\begin{array}{r} \frac{1}{2} \times 14 = 7 \\ 14 \times 2 = \underline{28} \\ 35\ \textit{Ans.} \end{array}$$

2. $18 \times 10\frac{2}{3}$.

PROCESS

$$\begin{array}{r} 18 \\ \times\, 10\frac{2}{3} \\ \hline 12 \\ \underline{180} \\ 192 \end{array}$$

EXPLANATION

$$\begin{array}{r} 18 \times \frac{2}{3} = 12 \\ 18 \times 10 = \underline{180} \\ 192 \end{array}$$

3. 24 by $12\frac{3}{4}$.

PROCESS

$$\begin{array}{r} 24 \\ \times\, 12\frac{3}{4} \\ \hline 18 \\ 48 \\ \underline{240} \\ 306 \end{array}$$

EXPLANATION

$$\begin{array}{r} 24 \times \frac{3}{4} = 24 \times \frac{1}{4} \times 3 = 18 \\ 24 \times 12 = 48 + 240 = \underline{288} \\ 306\ \textit{Ans.} \end{array}$$

Exercise

370. Use pencil only when necessary.

1. $1\frac{1}{2} \times 16$	2. $1\frac{1}{4} \times 16$	3. $1\frac{1}{8} \times 16$
4. $2\frac{1}{2} \times 8$	5. $2\frac{2}{3} \times 18$	6. $2\frac{1}{8} \times 8$
7. $4\frac{1}{3} \times 18$	8. $7\frac{1}{2} \times 20$	9. $2\frac{2}{3} \times 27$
10. $5\frac{1}{4} \times 8$	11. $3\frac{1}{4} \times 32$	12. $5\frac{1}{4} \times 40$
13. $6\frac{1}{4} \times \$4$	14. $2\frac{3}{4} \times 16$	15. $1\frac{3}{8} \times 24$
16. $2\frac{5}{8} \times 32$	17. $5\frac{3}{4} \times 20$	18. $7\frac{7}{8} \times 40$
19. $2\frac{3}{8} \times \$4$	20. $132 \times 27\frac{5}{6}$	21. $76 \times 36\frac{1}{4}$
22. $88 \times 18\frac{1}{8}$	23. $136 \times 19\frac{1}{2}$	24. $88 \times 29\frac{3}{8}$

25. From $\frac{1}{2}$ of 96 take $\frac{5}{8}$ of 72.

26. To $\frac{3}{4}$ of 48 add $\frac{3}{8}$ of 64.

27. From $\frac{5}{8}$ of 120 take $\frac{3}{4}$ of 36.

28. To $\frac{1}{4}$ of 72 add $\frac{3}{4}$ of 36.

Written Problems

371. 1. Two boys together earn \$ 24. One boy earned $\frac{1}{4}$ of the amount. How much did each earn?

2. A farmer bought a flock of 150 sheep at \$ $6\frac{1}{2}$ each. How much did he pay for the flock?

3. A merchant sold 48 T. of coal at \$ $6\frac{1}{4}$ per T. What was the amount of the sale?

4. There are $5\frac{1}{2}$ yd. in a rod. How many yd. in the side of a field that is 40 rods long?

5. A man pays $ $5\frac{1}{4}$ a week for board for 12 wk. How much did he pay?

6. A man gives $ $216\frac{1}{2}$ to his two sons. If one son receives $ 128, how much does the other son receive?

7. A boy earned $ 5.58 the first week and one half as much more the following week. How much did he earn the second week?

8. A room is 24 ft. long. It is $\frac{2}{3}$ as wide. How wide is it?

9. A grocer made up 150 packages of spices, each containing $\frac{2}{3}$ of a pound. How many pounds of spices did he have?

10. A florist set out 144 plants. During the winter $\frac{1}{6}$ of them died. How many lived?

Comparison — Oral

372. 1. Which is the greater, $\frac{1}{2}$ or $\frac{2}{3}$ of a unit?

2. Which is the greater, $\frac{2}{3}$ or $\frac{5}{6}$ of a unit?

3. Which is the greater, $\frac{1}{2}$ or $\frac{5}{6}$ of a unit?

4. How much greater is $\frac{5}{6}$ than $\frac{2}{3}$ of a unit?

5. How much greater is $\frac{2}{3}$ than $\frac{1}{2}$ of a unit?

6. How much greater is $\frac{5}{6}$ than $\frac{1}{2}$ of a unit?

7. How much greater is $\frac{1}{3}$ than $\frac{1}{6}$ of a unit?

8. How much less is $\frac{1}{4}$ of a unit than $\frac{1}{3}$ of a unit?

Written Exercise

373. Find the answer, illustrating by diagram.

1. $\frac{1}{2}+\frac{1}{4}=\frac{}{4}$	2. $\frac{1}{2}-\frac{1}{3}=\frac{}{6}$
3. $\frac{1}{3}+\frac{1}{6}=\frac{}{6}$	4. $1-\frac{2}{3}=\frac{}{3}$
5. $\frac{1}{2}+\frac{3}{4}=\frac{}{4}$	6. $\frac{1}{2}+\frac{3}{4}=1\frac{}{4}$
7. $\frac{1}{3}+\frac{5}{6}=\frac{}{6}$	8. $\frac{1}{3}+\frac{5}{6}=1\frac{}{6}$
9. $\frac{1}{2}+\frac{2}{3}=\frac{}{6}$	10. $\frac{1}{2}+\frac{2}{3}=1\frac{}{6}$
11. $1\frac{1}{6}-\frac{5}{6}=\frac{}{6}$	12. $1\frac{1}{6}-\frac{5}{6}=\frac{}{3}$
13. $1\frac{1}{3}-\frac{2}{3}=\frac{}{3}$	14. $1\frac{1}{2}-\frac{5}{6}=\frac{}{6}$
15. $1\frac{1}{2}-\frac{5}{6}=\frac{}{3}$	16. $1\frac{2}{3}-\frac{5}{6}=\frac{}{6}$

Addition — Oral

374. Add:

1. $\begin{array}{r}\frac{1}{2}\\ 1\frac{1}{2}\\ \hline\end{array}$	2. $\begin{array}{r}2\frac{1}{3}\\ \frac{1}{3}\\ \hline\end{array}$	3. $\begin{array}{r}3\frac{2}{3}\\ \frac{1}{3}\\ \hline\end{array}$	4. $\begin{array}{r}4\frac{1}{4}\\ \frac{1}{4}\\ \hline\end{array}$
5. $\begin{array}{r}1\frac{3}{4}\\ \frac{1}{4}\\ \hline\end{array}$	6. $\begin{array}{r}5\frac{1}{8}\\ \frac{1}{8}\\ \hline\end{array}$	7. $\begin{array}{r}8\frac{3}{8}\\ \frac{1}{8}\\ \hline\end{array}$	8. $\begin{array}{r}3\frac{1}{6}\\ \frac{1}{6}\\ \hline\end{array}$
9. $\begin{array}{r}\frac{5}{6}\\ 7\frac{1}{6}\\ \hline\end{array}$	10. $\begin{array}{r}\frac{1}{2}\\ 10\frac{1}{4}\\ \hline\end{array}$	11. $\begin{array}{r}\frac{1}{2}\\ \frac{3}{4}\\ \hline\end{array}$	12. $\begin{array}{r}\frac{1}{2}\\ \frac{1}{3}\\ \hline\end{array}$
13. $\begin{array}{r}\frac{1}{2}\\ \frac{2}{3}\\ \hline\end{array}$	14. $\begin{array}{r}\frac{1}{2}\\ \frac{1}{6}\\ \hline\end{array}$	15. $\begin{array}{r}\frac{1}{3}\\ \frac{1}{6}\\ \hline\end{array}$	16. $\begin{array}{r}\frac{2}{3}\\ \frac{1}{6}\\ \hline\end{array}$
17. $\begin{array}{r}\frac{2}{3}\\ \frac{5}{6}\\ \hline\end{array}$	18. $\begin{array}{r}\frac{2}{3}\\ \frac{1}{2}\\ \hline\end{array}$	19. $\begin{array}{r}\frac{3}{8}\\ \frac{1}{3}\\ \hline\end{array}$	20. $\begin{array}{r}\frac{3}{8}\\ \frac{1}{4}\\ \hline\end{array}$

Subtraction — Oral

375. Subtract:

1. $\frac{5}{6}$	2. $\frac{3}{4}$	3. $\frac{7}{8}$	4. $\frac{5}{6}$
$\underline{\frac{1}{3}}$	$\underline{\frac{1}{2}}$	$\underline{\frac{1}{2}}$	$\underline{\frac{2}{3}}$
5. $\frac{5}{8}$	6. $\frac{7}{8}$	7. $\frac{3}{4}$	8. $\frac{2}{3}$
$\underline{\frac{1}{4}}$	$\underline{\frac{3}{4}}$	$\underline{\frac{3}{8}}$	$\underline{\frac{1}{6}}$
9. 1	10. 1	11. 1	12. 1
$\underline{\frac{1}{3}}$	$\underline{\frac{1}{4}}$	$\underline{\frac{1}{6}}$	$\underline{\frac{1}{8}}$
13. 1	14. 1	15. 1	16. 1
$\underline{\frac{1}{2}}$	$\underline{\frac{3}{8}}$	$\underline{\frac{5}{6}}$	$\underline{\frac{3}{4}}$

Written Problems

376. 1. In a farm there are $27\frac{1}{2}$ A. in 1 field, $20\frac{1}{4}$ A. in another field, and $\frac{1}{8}$ A. in another. How many acres in the 3 fields?

2. A farmer sold $37\frac{1}{2}$ A. from his farm of 160 A. How many acres has he left?

3. A dealer bought $10\frac{1}{3}$ doz. eggs from one farmer, $22\frac{2}{3}$ doz from another. How many dozen did he buy?

4. A grocer sold $12\frac{1}{2}$ doz. eggs from his stock of 33 doz. eggs. How many dozen were left?

5. From a roll of ribbon containing $20\frac{1}{2}$ yd. a piece $7\frac{1}{3}$ yd. long was cut. How many yards remained?

6. A dressmaker made 2 waists. One of these waists required $2\frac{2}{3}$ yd. of silk and the other $3\frac{1}{6}$ yd. of silk. How many yards were used for both waists?

7. Find the cost of 4 books at \$ $1\frac{1}{4}$ each.

8. Find the cost of the following bill of goods:

4 yd. of ribbon at \$ $\frac{1}{2}$ per yard.
12 yd. of serge at \$ $1\frac{1}{4}$ per yard.
8 yd. of velvet at \$ $3\frac{3}{4}$ per yard.

9. Make out a bill for goods purchased from some dealer in your town.

10. Make a receipted bill for work and material furnished by William Edwards, New York City, to David Jones:

3 days' labor at \$ $3\frac{1}{2}$ per day.
Lumber furnished \$ $5\frac{1}{4}$.
Paint, etc. \$ $1\frac{3}{4}$.

To Find what Fraction One Number is of Another Number

377. 1. Draw a rectangle 4 in. long and $\frac{3}{4}$ as wide.

2. Divide the length into 4 equal parts. How long is each part?

3. Divide the width into 3 equal parts. How wide is each part?

4. Draw lines through these points to divide the rectangle into small squares.

5. How many small squares are there in the rectangle?

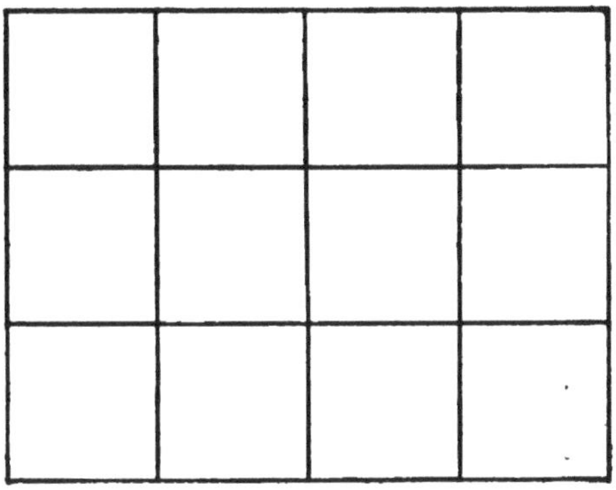

6. How many small squares in $\frac{1}{2}$ the rectangle? In $\frac{1}{4}$ the rectangle? In $\frac{1}{3}$ the rectangle?

7. 6 squares are what part of 12 squares?

8. 4 squares are what part of 12 squares?

9. 9 squares are what part of 12 squares?

10. What part of 12 squares are 3 squares? 8 squares? 10 squares? 1 square?

11. What part of 6 squares are 6 squares? 3 squares? 2 squares? 4 squares?

Oral Exercise

378. 1. What part of 1 ft. is 1 in.?

2. 2 in. is what part of 12 in.?

3. What part of 1 ft. is 3 in.? 4 in.? 6 in.? 8 in.? 9 in.? 10 in.?

4. What part of 1 doz. is 2 units? 3 units? 4 units? 6 units? 8 units?

5. What part of 1 lb. is 4 oz.? 8 oz.? 12 oz.? 2 oz.? 1 oz.?

6. What part of 1 gal. is 1 pt.? 1 qt.? 2 qt.? 3 qt.? 4 qt.?

7. What part of 1 bu. is 1 pk.? 2 pk.? 3 pk.?

8. What part of 1 hr. is 15 min.? 20 min.? 30 min.? 40 min.? 10 min.?

9. What part of $1 is 10¢? 20¢? 25¢? 50¢? 75¢?

10. What part of 1 wk. is 1 da.? 2 da.? 3 da.? 5 da.? 7 da.?

Oral Exercise

379. What part:

1. Of 8 is 4?	2. Of 12 is 4?	3. Of 20 is 4?
4. Of 9 is 3?	5. Of 12 is 3?	6. Of 15 is 3?
7. Of 10 is 5?	8. Of 15 is 5?	9. Of 25 is 5?

Written Problems—Review

380. 1. If 1 doz. oranges cost 24¢, what will $\frac{2}{3}$ of a dozen cost?

2. At 24¢ a dozen, what will be the cost of $1\frac{2}{3}$ doz.?

3. A piece of goods is worth $75. What is $\frac{3}{5}$ of the piece worth?

4. A boy invited 16 playmates to a picnic. If $\frac{3}{4}$ of them accepted, how many boys attended the picnic?

5. A man planted 36 trees. If $\frac{5}{6}$ of the number lived, how many died?

6. Mr. Jones sold $\frac{2}{3}$ of a piece of land which he valued at $ 87. What did he receive from the sale?

7. A girl has read $\frac{2}{3}$ of a book of 165 pages. How many pages has she read?

8. If a girl has read $\frac{2}{3}$ of a book of 165 pages, how many pages has she still to read?

9. A piece of ribbon, 4 yd. in length, is cut into 3 equal pieces. What is the length of each piece?

Review Problems — Miscellaneous

381. 1. If hay costs $ 12 a ton, what is the cost of 47 tons?

2. If hay costs $ 12 a ton, what is the cost of $1\frac{1}{3}$ tons?

3. What is the cost of 15 doz. eggs at 39¢ per dozen?

4. Yesterday there were 14 pupils absent from class. If this number is just $\frac{1}{6}$ of the class, how many pupils are there in the class?

5. How many dozens are there in one hundred thirty-two oranges?

6. A merchant buys sugar at 6¢ a pound and sells it for 8¢ a pound. What is his profit on 46 lb.?

7. A girl has 136 beads, which is 4 times as many as her schoolmate has. How many has her schoolmate?

8. The sum of 2 numbers is 237. One of the numbers is 94. What is the other?

9. A woman paid $11.25 for 9 yd. of cloth. What was the price of 1 yd.?

10. At $\$\frac{1}{3}$ per yard, what is the cost of 75 yd. of mull?

11. A boy can write 25 words a minute. How many can he write in half an hour?

Reduction of Fractions

382. Copy and complete:

1. $\frac{1}{2}=\frac{}{4}$	2. $\frac{1}{3}=\frac{}{6}$	3. $\frac{2}{3}=\frac{}{6}$
4. $1=\frac{}{2}$	5. $1=\frac{}{3}$	6. $1=\frac{}{4}$
7. $\frac{2}{3}=\frac{}{6}$	8. $\frac{3}{6}=\frac{}{2}$	9. $\frac{2}{4}=\frac{}{2}$

In this exercise we have been changing the form of the fraction without changing the value. The process of changing the form of the fraction without changing its value is called **reduction.**

Exercise in Construction

383. Draw 4 squares of equal size.

Divide the sides and draw lines to divide each square into 4 equal parts.

Let the square represent 1 unit.

 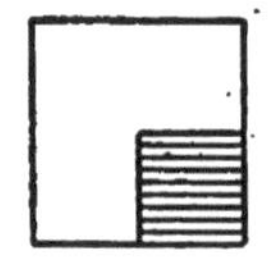 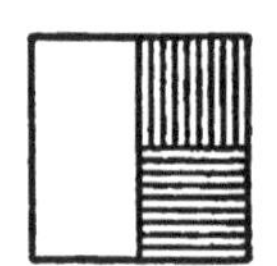 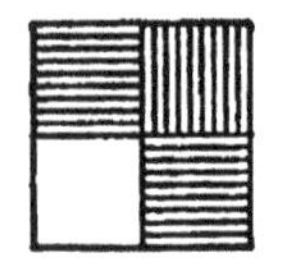 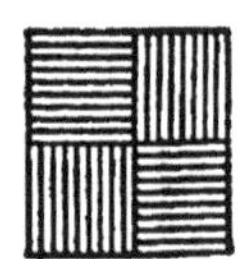

1. How many fourths in 1 unit?
2. How many fourths in $1\frac{1}{4}$ units?

3. How many fourths in $1\frac{2}{4}$ units?
4. How many fourths in 2 units?
5. How many fourths in 4 units?
6. How many fourths in 3 units? In 5 units?
7. How many halves in 1 unit? In 2 units?
8. How many halves in $1\frac{1}{2}$ units?
9. How many thirds in 1 unit? In 2 units?
10. How many thirds in 3 units? In 7 units?
11. How many thirds in $1\frac{1}{3}$ units? In $2\frac{2}{3}$ units?
12. How many sixths in 1 unit? In $2\frac{5}{6}$ units?

Fractions like $\frac{1}{2}$, $\frac{2}{4}$, $\frac{1}{3}$, $\frac{3}{4}$, in which the value is less than one, are called **proper fractions.**

Fractions like $\frac{2}{2}$, $\frac{6}{4}$, $\frac{4}{3}$, in which the value is equal to or greater than one, are called **improper fractions.**

The numerator of a proper fraction is less than the denominator.

The numerator of an improper fraction is equal to or greater than the denominator.

Written Exercise

384. Copy and place in separate columns the proper and improper fractions:

1.	$\frac{2}{3}$	$\frac{4}{7}$	$\frac{9}{5}$	$\frac{8}{8}$	$\frac{3}{2}$
2.	$\frac{4}{5}$	$\frac{6}{7}$	$\frac{7}{5}$	$\frac{7}{7}$	$\frac{4}{4}$
3.	$\frac{9}{8}$	$\frac{9}{10}$	$\frac{10}{9}$	$\frac{5}{2}$	$\frac{2}{5}$
4.	$\frac{15}{16}$	$\frac{5}{3}$	$\frac{5}{4}$	$\frac{5}{6}$	$\frac{17}{16}$

To Change a Whole Number to an Improper Fraction

385. 1. How many fourths in 3?

PROCESS

$1 = \frac{4}{4}$

$3 = \frac{3 \times 4}{4} = \frac{12}{4}$ *Ans.*

EXPLANATION

There are $\frac{4}{4}$ in 1.
There are 3 times $\frac{4}{4}$ in 3.

2. How many fourths in 2? In 4? In 5?
3. How many thirds in 3? In 5? In 8?
4. How many fifths in 2? In 5? In 10?
5. How many eighths in 4? In 7? In 12?
6. How many tenths in 3? In 10? In 15?

Written Exercise

386. 1. How many thirds in 275?

PROCESS

$1 = \frac{3}{3}$

$275 = \frac{275 \times 3}{3} = \frac{825}{3}$ *Ans.*

Write the explanation.

2. How many thirds in 25? In 11? In 20?
3. How many fourths in 25? In 50? In 100?
4. How many sixths in 12? In 6? In 8?
5. How many eighths in 4? In 8? In 12?
6. How many fourths in $4\frac{1}{2}$? In $10\frac{1}{4}$? In $12\frac{1}{2}$?
7. How many sixths in $5\frac{1}{6}$? In $6\frac{5}{6}$? In $16\frac{2}{3}$?

To Change a Mixed Number to an Improper Fraction

387. 1. How many fifths in $2\frac{3}{5}$?

PROCESS

$1 = \frac{5}{5}$

$2 = \frac{10}{5}$

$2\frac{3}{5} = \frac{10}{5} + \frac{3}{5} = \frac{13}{5}$ *Ans.*

EXPLANATION

$2\frac{3}{5}$ means $2 + \frac{3}{5}$

$2 = \frac{10}{5}$

$2\frac{3}{5} = \frac{10}{5} + \frac{3}{5} = \frac{13}{5}$

2. How many fifths in $5\frac{2}{5}$? In $4\frac{4}{5}$?
3. How many sixths in $6\frac{1}{6}$? In $8\frac{5}{6}$?
4. How many sevenths in $4\frac{3}{7}$? In $10\frac{3}{7}$?
5. How many eighths in $12\frac{3}{8}$? In $15\frac{5}{8}$?
6. How many tenths in $10\frac{1}{10}$? In $5\frac{2}{10}$?

Written Exercise

388. Reduce to an improper fraction.

1. $75\frac{3}{4}$

PROCESS

$$\begin{array}{r} 75 \\ 4 \\ \hline 3 \\ 300 \\ \hline 303 \end{array}$$

303 fourths $= \frac{303}{4}$ *Ans.*

EXPLANATION

Reduce $75\frac{3}{4}$ to fourths.

$75\frac{3}{4}$ means $75 + \frac{3}{4}$

$75 = \frac{75 \times 4}{4}$

$75\frac{3}{4} = \frac{75 \times 4}{4} + \frac{3}{4} = \frac{303}{4}$

2. $20\frac{1}{4}$	3. $20\frac{1}{5}$	4. $40\frac{7}{8}$	5. $12\frac{3}{7}$
6. $12\frac{1}{8}$	7. $33\frac{1}{3}$	8. $66\frac{2}{3}$	9. $16\frac{2}{3}$
10. $15\frac{3}{8}$	11. $21\frac{1}{2}$	12. $34\frac{3}{4}$	13. $51\frac{3}{5}$
14. $152\frac{5}{8}$	15. $241\frac{1}{12}$	16. $353\frac{7}{12}$	17. $425\frac{4}{5}$

Oral Problems

389. 1. How many pieces of lace, each $\frac{1}{4}$ yd. long, can be cut from a piece 12 yd. long?

2. How many boxes, each holding $\frac{1}{2}$ doz., can be filled from a crate containing 6 doz. eggs?

3. How many pecks of corn in a sack containing $2\frac{1}{4}$ bu.?

4. How many packages of tea, each containing $\frac{1}{2}$ lb., can be put up from a can containing 10 lb.?

5. How many 10-cent pieces in \$ $5\frac{1}{10}$?

To Reduce an Improper Fraction to a Whole Number or Mixed Number

390. 1. Reduce $\frac{24}{4}$ to a whole number.

PROCESS BY CANCELATION

$$\frac{24}{4} = \frac{\overset{6}{\cancel{24}}}{\underset{1}{\cancel{4}}} = 6.$$

EXPLANATION. — Cancel the common factor 4 from the numerator and denominator.

2. Reduce $\frac{159}{14}$ to a whole number.

PROCESS BY DIVISION

```
    11 5/14 Ans.
14)159
   14
    19
    14
     5
```

EXPLANATION.— Since there are no common factors, reduce by division.

A fraction is an indicated division. The numerator is the dividend, the denominator is the divisor.

IMPORTANT PRINCIPLE IN FRACTIONS

I

Dividing both terms of a fraction by the same number does not change the value of the fraction.

Written Exercises

391. Write as whole or mixed numbers:

1. $\frac{36}{6}$	2. $\frac{42}{7}$	3. $\frac{48}{8}$	4. $\frac{54}{9}$
5. $\frac{60}{7}$	6. $\frac{14}{21}$	7. $\frac{80}{9}$	8. $\frac{101}{10}$
9. $\frac{140}{12}$	10. $\frac{265}{9}$	11. $\frac{335}{11}$	12. $\frac{421}{12}$

To Reduce a Fraction to Lower Terms

392. 1. Reduce $\frac{36}{48}$ to lower terms.

PROCESS

$$\frac{36}{48}=\frac{\overset{1}{\not 6}\times\overset{3}{\not 6}}{\underset{1}{\not 6}\times\underset{4}{\not 8}}=\frac{3}{4}$$

EXPLANATION. — Cancel the common factors in the numerator and the denominator.

2. Reduce $\frac{540}{720}$ to lower terms:

PROCESS

$$\frac{540}{720}=\frac{\overset{1}{\not{10}}\times\overset{3}{\not 9}\times\overset{1}{\not 6}}{\underset{1}{\not{10}}\times\underset{4}{\not{12}}\times\underset{1}{\not 6}}=\frac{3}{4}$$

EXPLANATION. — Cancel the common factors from the numerator and denominator.

When all the common factors have been canceled from the numerator and the denominator, the fraction is in its lowest terms.

In practice the factor 1 is not written.

Exercise

393. Reduce to lowest terms:

1. $\frac{9}{12}$	2. $\frac{8}{12}$	3. $\frac{6}{24}$	4. $\frac{9}{18}$	5. $\frac{7}{21}$
6. $\frac{14}{21}$	7. $\frac{12}{20}$	8. $\frac{12}{18}$	9. $\frac{9}{27}$	10. $\frac{12}{24}$
11. $\frac{9}{24}$	12. $\frac{16}{20}$	13. $\frac{20}{30}$	14. $\frac{30}{40}$	15. $\frac{11}{33}$
16. $\frac{10}{25}$	17. $\frac{10}{15}$	18. $\frac{10}{35}$	19. $\frac{14}{56}$	20. $\frac{24}{32}$

To Reduce a Fraction to Higher Terms

394. 1. Change $\frac{3}{4}$ to eighths.

PROCESS

$$\frac{3}{4}=\frac{3\times 2}{4\times 2}=\frac{6}{8}$$

EXPLANATION. — To change fourths to eighths, make the denominator of the new fraction 8. To make the denominator 8, multiply both terms by 2.

2. Change $2\frac{1}{5}$ to tenths.

PROCESS

$$2\tfrac{1}{5}=\tfrac{11}{5}$$

$$\frac{11}{5}=\frac{11\times 2}{5\times 2}=\frac{22}{10}$$

EXPLANATION. — Reduce the mixed number to an improper fraction.

To change fifths to tenths, make the denominator of the new fraction tenths.

Then multiply both terms by the factor that will change the denominator to the required denominator.

AN IMPORTANT PRINCIPLE IN FRACTIONS

II

Multiplying both terms of a fraction by the same number does not change the value of the fraction.

Written Exercise

395. Reduce:

1. $\frac{1}{2}$ $\frac{2}{3}$ $\frac{3}{3}$ $1\frac{1}{3}$ $2\frac{2}{3}$ to sixths.
2. $\frac{1}{2}$ $\frac{1}{4}$ $\frac{3}{4}$ $1\frac{1}{2}$ $2\frac{1}{4}$ to eighths.
3. $\frac{1}{2}$ $\frac{1}{3}$ $\frac{2}{3}$ $\frac{3}{4}$ $4\frac{2}{3}$ to twelfths.
4. $\frac{1}{2}$ $\frac{1}{5}$ $\frac{2}{5}$ $\frac{3}{5}$ $\frac{4}{5}$ to tenths.
5. $1\frac{1}{2}$ $3\frac{1}{2}$ $1\frac{2}{5}$ $3\frac{2}{5}$ $5\frac{1}{5}$ to tenths.
6. $\frac{1}{8}$ $\frac{3}{4}$ $\frac{3}{8}$ $\frac{1}{4}$ $\frac{5}{8}$ to sixteenths.
7. $4\frac{1}{8}$ $12\frac{1}{4}$ $5\frac{5}{8}$ $16\frac{1}{2}$ $10\frac{7}{8}$ to sixteenths.
8. $\frac{1}{2}$ $\frac{1}{4}$ $\frac{3}{4}$ $\frac{7}{8}$ $\frac{3}{2}$ to eighths.
9. $5\frac{4}{5}$ $10\frac{2}{5}$ $4\frac{1}{2}$ $3\frac{3}{5}$ $1\frac{1}{10}$ to tenths.
10. $\frac{1}{10}$ $\frac{2}{10}$ $\frac{3}{10}$ $\frac{4}{10}$ $\frac{5}{10}$ to hundredths.
11. $\frac{6}{10}$ $\frac{7}{10}$ $\frac{8}{10}$ $\frac{9}{10}$ $\frac{10}{10}$ to hundredths.
12. $\frac{1}{2}$ $\frac{1}{5}$ $\frac{3}{5}$ $\frac{1}{4}$ $\frac{3}{4}$ to hundredths.
13. $\frac{1}{20}$ $\frac{7}{20}$ $\frac{5}{20}$ $\frac{8}{20}$ to hundredths.
14. $\frac{4}{25}$ $\frac{3}{25}$ $\frac{16}{25}$ $\frac{10}{50}$ $\frac{20}{50}$ to hundredths.

Similar Fractions and Common Denominators

396. $\frac{1}{8}$, $\frac{5}{8}$, $\frac{7}{8}$ are similar fractions. Fractions that have the same denominator are called *similar fractions*.

1. Reduce $\frac{1}{2}$, $\frac{2}{3}$ and $\frac{3}{4}$ to similar fractions, that is, to fractions having the same denominator.

PROCESS

$\frac{1}{2}=\frac{6}{12}$

$\frac{2}{3}=\frac{8}{12}$

$\frac{3}{4}=\frac{9}{12}$

EXPLANATION. — By inspection we find that these fractions may all be reduced to twelfths.

$\frac{1}{2}=\frac{6}{12}$, $\frac{2}{3}=\frac{8}{12}$, $\frac{3}{4}=\frac{9}{12}$

$\frac{6}{12}$, $\frac{8}{12}$, and $\frac{9}{12}$ are similar fractions. The common denominator of these fractions is 12.

What is the sum of 6 twelfths, 8 twelfths, and 9 twelfths?

2. Reduce $1\frac{1}{2}$, $\frac{7}{10}$, $\frac{4}{5}$ to similar fractions.

PROCESS

$1\frac{1}{2} = \frac{3}{2} = \frac{15}{10}$

$\frac{7}{10} = \frac{7}{10}$

$\frac{4}{5} = \frac{8}{10}$

EXPLANATION.—Reduce the mixed number, $1\frac{1}{2}$, to the improper fraction, $\frac{3}{2}$. Find by inspection or by factoring the least common denominator (l. c. d.). The l. c. d. = 10.

The fractions must be reduced to tenths: $1\frac{1}{2} = \frac{15}{10}$, $\frac{7}{10} = \frac{7}{10}$, $\frac{4}{5} = \frac{8}{10}$.

Tell how to add similar fractions.

$$\frac{15}{10} + \frac{7}{10} + \frac{8}{10} = \frac{?}{10} = 3$$

3. Find the sum of $\frac{4}{5}$, $\frac{3}{4}$, and $\frac{3}{10}$ by trial or inspection.

$\frac{3}{10}$

$\frac{4}{5}$

$\frac{3}{4}$

EXPLANATION.—10 is the largest denominator. It may be the common denominator.

The denominators 10 and 5 are exact divisors of 10.

But the denominator 4 is not an exact divisor of 10.

For that reason, 10 is not a common denominator.

$4 = 2 \times 2$.

Try 10×2.

All three denominators 4, 5, and 10 are exact divisors of 20.

For that reason, 20 is a common denominator.

$\frac{4}{5}$, $\frac{3}{4}$, and $\frac{3}{10}$ reduced to similar fractions are $\frac{16}{20}$, $\frac{15}{20}$, and $\frac{6}{20}$. Their sum is $\frac{37}{20} = 1\frac{17}{20}$.

Subtract:

4. $1\frac{1}{5} - \frac{3}{4}$.

PROCESS

$$1\frac{1}{5} = \frac{6}{5} = \frac{24}{20}$$
$$\frac{3}{4} = \frac{15}{20}$$
$$\frac{9}{20} \quad \textit{Ans.}$$

EXPLANATION.—Reduce the mixed number to an improper fraction. Then reduce the fractions to similar fractions. Subtract.

Written Exercises

397. Add or subtract:

1. $1\frac{1}{2}+\frac{3}{4}+\frac{1}{8}$	2. $1\frac{1}{3}+\frac{2}{3}+\frac{1}{6}$	3. $1\frac{1}{4}+\frac{3}{4}+\frac{3}{8}$
4. $2\frac{1}{2}+\frac{1}{3}+\frac{5}{12}$	5. $\frac{1}{2}+\frac{1}{3}+\frac{3}{4}$	6. $\frac{4}{5}+\frac{1}{2}+\frac{3}{10}$
7. $\frac{3}{10}+\frac{1}{2}+\frac{4}{5}$	8. $1\frac{1}{8}+\frac{7}{8}+\frac{1}{4}$	9. $\frac{5}{4}-\frac{1}{4}-\frac{3}{4}$
10. $1-\frac{1}{4}-\frac{1}{3}$	11. $1-\frac{1}{2}-\frac{1}{5}$	12. $1-\frac{1}{8}-\frac{3}{4}$
13. $\frac{2}{3}-\frac{1}{2}+\frac{1}{4}$	14. $\frac{3}{2}-\frac{2}{5}+\frac{1}{4}$	15. $1\frac{5}{6}-\frac{1}{3}-\frac{1}{2}$

Written Problems

398. 1. A boy eats $\frac{1}{2}$ of an apple and his sister $\frac{1}{3}$ of it. What part of the apple is left?

2. A girl has $1\frac{1}{4}$ pounds of candy. She eats $\frac{1}{3}$ of a pound. What part of a pound has she left?

3. John spends $\$2\frac{1}{2}$ for a railroad ticket and $\$\frac{1}{4}$ for a meal. How much does he spend in all?

4. Mary is reading a book. On Monday she reads $\frac{1}{2}$ of it, on Tuesday $\frac{1}{4}$, and on Wednesday $\frac{1}{8}$ of it. How much of the book has she still to read?

5. If the book has 400 pages, how many pages did Mary read on Monday? On Tuesday? On Wednesday?

6. What is the weight of three packages if the first package weighs $\frac{1}{8}$ of a pound, the second $\frac{1}{4}$ of a pound, the third $1\frac{1}{2}$ pounds?

7. A tailor uses $4\frac{1}{2}$ yd. of cloth for a coat; $1\frac{1}{3}$ yd. for a vest, and $3\frac{1}{4}$ yd. for a pair of trousers. How many yards does he use for the suit?

8. A man had $12\frac{1}{2}$ gal. of milk in a can and adds $5\frac{1}{4}$ gal. more. How many gallons are there in the can?

9. A farmer sowed $37\frac{3}{4}$ acres with wheat and later $12\frac{1}{2}$ acres more. How many had he then sowed?

10. A playground is in the form of a rectangle $256\frac{3}{8}$ ft. long and $178\frac{3}{4}$ ft. wide. How long is the fence that encloses the playground?

11. From a barrel containing $25\frac{2}{5}$ gallons of vinegar a grocer sells $13\frac{1}{10}$ gallons. How many gallons are left?

12. I buy butter for $\$1\frac{3}{4}$, sugar for $\$\frac{7}{8}$, and rolls for $\$\frac{1}{4}$. How much change do I get, if I pay with a $5 bill?

13. One man has $48, another has $\frac{2}{3}$ as much, a third has $\frac{1}{4}$ as much. How many dollars do they all have?

14. A jar containing butter weighs $24\frac{7}{8}$ pounds. The empty jar is known to weigh $4\frac{3}{4}$ pounds. Find the weight of the butter.

15. A rectangular picture frame is $18\frac{1}{2}$ in. long and $12\frac{3}{4}$ in. wide. What is the perimeter of the frame?

16. The sum of two numbers is $20\frac{2}{3}$; one of them is $5\frac{1}{6}$. What is the other?

17. At $8 a ton, find the cost of $9\frac{3}{4}$ tons of hay.

18. A man bought a house for $ 3500, and paid $\frac{3}{5}$ of the price at the time of purchase. How much remained to be paid?

19. At $$6\frac{1}{3}$ a cord, find the cost of 9 cords of wood.

Find the cost of

20. $10\frac{1}{2}$ lb. of rice at 8¢ a pound.

21. 10 lb. of sugar at $5\frac{1}{2}$¢ a pound.

22. $12\frac{1}{4}$ lb. of coffee at 36¢ a pound.

23. A clerk earns $25 a week and spends $\frac{3}{5}$ of it. How much does he spend?

To Find what Fraction One Number is of Another

399. 1. What part of 16 is 4?

PROCESS

$\frac{4}{16} = \frac{1}{4}$

EXPLANATION. — The part that 4 units are of 16 units is indicated by the fraction $\frac{4}{16}$ or $\frac{1}{4}$.

2. A farmer had 40 head of cattle. He sold 8 head. What part did he sell?

PROCESS	EXPLANATION. — The number of
	cattle $= 40$
$\frac{8}{40} = \frac{1}{5}$	The number sold $= 8$
	The part sold $= \frac{8}{40} = \frac{1}{5}$

Oral Problems

400. 1. A book contains 200 pages. Mary has read 100 pages. What part of the book has she read?

2. A man earned $ 60 and spent $ 20. What part of his earnings did he spend?

3. A garden is 150 ft. long and 60 ft. wide. What part of the length is the width?

4. A merchant paid $ 300 for goods and sold them for $ 400. What part of the cost is the selling price?

5. What part of a dollar is a dime?

6. What part of a foot is 2 inches? 3 inches?

Written Review

401. Add:

1.	$61\frac{3}{4}$	$27\frac{1}{4}$	$28\frac{1}{2}$	$75\frac{3}{4}$	$101\frac{1}{6}$
	$79\frac{1}{4}$	$26\frac{1}{2}$	$82\frac{1}{3}$	$64\frac{1}{8}$	$97\frac{1}{2}$
2.	$191\frac{1}{4}$	$265\frac{2}{3}$	$31\frac{3}{8}$	$39\frac{1}{2}$	612
	$71\frac{1}{4}$	$67\frac{1}{6}$	$121\frac{1}{4}$	$171\frac{1}{2}$	$171\frac{4}{5}$

Subtract:

3.	$171\frac{3}{4}$	$685\frac{2}{3}$	$789\frac{5}{6}$	$97\frac{7}{8}$	$75\frac{7}{8}$
	$97\frac{1}{4}$	$173\frac{1}{6}$	$543\frac{1}{3}$	$79\frac{3}{4}$	$37\frac{1}{2}$
4.	$197\frac{1}{2}$	$79\frac{1}{2}$	$191\frac{1}{4}$	$700\frac{1}{3}$	$798\frac{1}{4}$
	$108\frac{1}{3}$	46	98	$93\frac{1}{6}$	$354\frac{1}{8}$

Written Review Problems

402. 1. In one day a boy carns $\$1\frac{1}{4}$ and the next day $\$2\frac{1}{2}$. What does he earn in the two days?

2. At $9 a ton, how many tons of hay can be bought for $819?

3. How many square feet in 1 square yard? How many square yards in 162 sq. ft.?

4. At 40 miles an hour, in what time does a train run 320 miles?

5. What is the cost of 2 bushels of nuts at 72¢ a peck?

6. What is the cost of 24 acres of land at $45.75 an acre?

7. What is the cost of building 7 miles of railroad at $14,560 a mile?

8. If 3 pieces of cloth cost $93, what is the cost of 1 piece? Of 8 pieces?

9. If 4 yd. of ribbon cost 36¢, what will 1 yd. cost? What will $17\frac{1}{3}$ yd. cost?

10. If 2 baskets of peaches cost 50¢, what will 13 baskets cost?

11. If 3 lb. of pork cost 36¢, what is the cost of 1 lb.? Of 23 lb.?

12. What is the cost of $5\frac{1}{2}$ lb. of coffee at 30¢ a pound?

13. If 5 towels sell for $1, for what do 15 towels sell?

14. If 2 suits of clothes can be had for $26, what will be the cost of 7 suits of the same quality?

15. A boy entering school pays for an arithmetic 65¢, for a notebook 15¢, for a geography $1.10. How much money must he receive in change, if he gives a $5 bill?

16. What is the cost of 6 tons of coal at $5.25 a ton?

17. What are the wages of 8 men per day at $2.75 each?

18. A party orders 16 plates of ice cream. What is their bill at 15¢ a plate?

19. A street car conductor collects 55 5-cent fares. How many dollars and cents does he collect?

20. Mary draws a line $\frac{2}{3}$ of a foot long and another $\frac{1}{2}$ of a foot long. How much longer is the first than the second?

21. How long are the two lines taken together?

22. Two boys start from the same place and run in opposite directions; one goes $\frac{2}{3}$ of a mile, the other $\frac{5}{6}$ of a mile. How far apart are they?

DECIMAL FRACTIONS

Oral Exercise

403. 1. In $1.37 the 3 stands for $\frac{?}{10}$ of a dollar?

2. In $1.37 the 7 stands for $\frac{?}{100}$ of a dollar?

3. How many tenths in a unit?

4. How many hundredths in a unit?

5. How many thousandths in a unit?

6. How many hundredths in one tenth?

7. How many thousandths in one tenth?

8. How many thousandths in one hundredth?

Written Exercise

404. Write in figures as common fractions:

1. One tenth. 2. Two tenths. 3. Three tenths.

4. One hundredth. 5. Three hundredths.

6. Twenty hundredths. 7. Seventy-five hundredths.

8. Five thousandths. 9. Nineteen thousandths.

10. One hundred ninety-nine thousandths.

405. Write in words:

1. $\frac{1}{10}$ 2. $\frac{7}{10}$ 3. $\frac{9}{10}$ 4. $\frac{13}{100}$ 5. $\frac{19}{100}$ 6. $\frac{29}{100}$

7. $\frac{50}{100}$ 8. $\frac{80}{100}$ 9. $\frac{175}{1000}$ 10. $\frac{123}{100}$ 11. $\frac{723}{1000}$

A fraction whose denominator is 10, 100, or 1000 is called a **decimal fraction**.

A decimal fraction may be written in three ways:

In words, as, one tenth, five tenths.

As a common fraction, as, $\frac{3}{10}$, $\frac{19}{100}$.

By using the decimal point, .3, .9, $ 1.25.

Oral Exercise

406. Read and compare the following decimal fractions:

In Words	As Common Fraction	With Decimal Point
One tenth	$\frac{1}{10}$	.1
Five tenths	$\frac{5}{10}$	.5
Nine tenths	$\frac{9}{10}$	.9
One hundredth	$\frac{1}{100}$	.01
Nineteen hundredths	$\frac{19}{100}$	.19
Ninety-one hundredths	$\frac{91}{100}$	.91
One thousandth	$\frac{1}{1000}$	.001
Nine thousandths	$\frac{9}{1000}$	.009
Twenty-nine thousandths	$\frac{29}{1000}$	.029
One hundred seventy-five thousandths	$\frac{175}{1000}$	.175

Oral Exercise

407. Read:

1.	.1	.5	.7	.9	.4	.3	.2	.8
2.	.01	.19	.25	.78	.50	.06	.20	.99
3.	.001	.011	.013	.019	.051	.125	.735	.999

When the meaning of a decimal fraction is understood, it is allowable, for brevity, to read decimal fractions as follows:

.6 Point, six
.17 Point, one, seven.

4. Give the shortened readings of the fractions in Ex. 3.

Written Exercise

408. Write, using the decimal point:

1. $\frac{4}{10}$	2. $\frac{4}{100}$	3. $\frac{4}{1000}$	4. $\frac{1}{10}$
5. $\frac{9}{10}$	6. $\frac{25}{100}$	7. $\frac{75}{1000}$	8. $\frac{475}{1000}$

Written Exercise

409. Write as a common fraction:

1. .5	2. .04	3. .125	4. .256
5. .8	6. .12	7. .005	8. .075
9. .6	10. .72	11. .625	12. .004
13. .1	14. .10	15. .500	16. .800

Written Exercise

410. Reduce to common fractions in lowest terms:

1. .5	2. .8	3. .75	4. .125	5. .625
6. .05	7. .08	8. .075	9. .025	10. .005
11. .02	12. .002	13. .0625	14. .050	

411. **Roman Numerals**

1. Read the numerals on the watch.

2. You see that

I	= 1	VII	= 7
II	= 2	VIII	= 8
III	= 3	IX	= 9
IIII or IV	= 4	X	= 10
V	= 5	XI	= 11
VI	= 6	XII	= 12

3. Remember that V = 5, X = 10.

4. Notice also that

IV = 5 − 1 = 4, VI = 5 + 1 = 6
IX = 10 − 1 = 9, XI = 10 + 1 = 11

5. We have

XIII = 10 + 3 = 13, XV = 10 + 5 = 15
XX = 10 + 10 = 20, XXI = 10 + 10 + 1 = 21

Read the following:

6. V X IV IX VI XI
7. VII XII XIII XIV XVI XVII
8. III VIII XVIII XX XIX XXI

Write in Roman numerals:

9.	5	6	4	9	10
10.	11	15	14	13	12
11.	20	19	21	22	23
12.	24	25	18	17	16

In the Roman notation, L = 50, C = 100, XL = 40, LX = 60, XC = 90, CX = 110, CC = 200, D = 500, M = 1000.

13. Read: LI, LV, XXX, XLI, CL, CLX, LXX, LXXX, LXI, LXV, CXL.

GENERAL REVIEW

Written Exercise

412. Add and check:

1.	2.	3.	4.
316	512	308	24
490	613	876	386
308	741	985	405
741	899	470	385
74	708	213	219
796	66	77	926

5.	6.	7.	8.
301	401	509	317
875	90	636	146
361	888	747	879
478	74	990	478
365	643	234	654
432	755	123	435

Subtract:

9.	10.	11.
3249	7986	4326
− 1986	− 5409	− 2789

12.	13.	14.
4590	1894	4507
− 1943	− 909	− 2345

15.	16.	17.
9876	87006	7000
− 8105	− 67805	− 2397

Add and check:

18.	19.	20.	21.
$ 1.34	$ 9.05	$ 10.25	$ 7.95
1.45	3.75	8.45	4.65
2.90	4.25	3.65	7.30

Subtract and check:

22.	23.	24.	25.
$ 145.63	$ 109.40	$ 760.09	$ 139.68
− 74.75	− 74.83	− 96.41	− 89.89

Multiply and check:

26. $14 \times \$ 7.65$.
27. $27 \times \$ 2.05$.
28. $31 \times \$ 4.09$.
29. $58 \times \$ 1.95$.

Divide and check:

30. 9)$ 132.75.
31. 27)$ 15.84.
32. 91)$ 99.19.
33. 39)$ 48.75.
34. 123×45.
35. 456×78.
36. 789×90.
37. 901×83.
38. If one factor of 221 is 13, find another factor.
39. If one factor of 1728 is 36, find another factor.
40. If one factor of 8526 is 98, find another factor.
41. If one factor of 299 is 23, find another factor.
42. 2385 has 53 as a factor, find another.
43. 1608 has 67 as a factor, find another.
44. Write down all the factors of 24.

TABLES FOR REFERENCE

United States Money

10 mills	= 1 cent (¢)
10 cents	= 1 dime (d.)
10 dimes	= 1 dollar ($)

Linear Measure

12 inches (in.)	= 1 foot (ft.)
3 feet	= 1 yard (yd.)
5½ yards	= 1 rod (rd.)
5,280 feet	= 1 mile (mi.)

Square Measure

144 square inches (sq. in.)	= 1 square foot (sq. ft.)
9 square feet	= 1 square yard (sq. yd.)
30¼ square yards	= 1 square rod (sq. rd.)
160 square rods	= 1 acre (A.)
640 acres	= 1 square mile (sq. mi.)

Cubic Measure

1,728 cubic inches (cu. in.)	= 1 cubic foot (cu. ft.)
27 cubic feet	= 1 cubic yard (cu. yd.)

Liquid Measure

2 pints (pt.)	= 1 quart (qt.)
4 quarts	= 1 gallon (gal.)

Dry Measure

2 pints	= 1 quart
8 quarts	= 1 peck (pk.)
4 pecks	= 1 bushel (bu.)

Avoirdupois

16 ounces (oz.)	= 1 pound (lb.)
2,000 pounds	= 1 ton (T.)

Time

60 seconds (sec.)	= 1 minute (min.)
60 minutes	= 1 hour (hr.)
24 hours	= 1 day (da.)
7 days	= 1 week (wk.)
12 months (mo.)	= 1 year (yr.)

Counting

12 things	= 1 dozen (doz.)
20 things	= 1 score
24 sheets	= 1 quire

TESTS OF ARITHMETICAL ABILITY FOR PUPILS OF THE THIRD AND FOURTH GRADES

Written tests like the ones proposed below, but with changes in the numbers and also slight changes in the wording, should be given, without previous notice, limiting the time to 12 or 15 minutes and requiring pupils to work the tests in the given order. Assign more questions than pupils are able to answer in the given time. One examination may be on Part I, another on Part II. As in the Courtes tests, either or both of two systems of marking may be adopted, (1) by marking each example 1 or 0, according as it is right or wrong, or (2) marking each example on the basis of the total number of steps in addition, subtraction, multiplication, or division involved. In Part I, question 2 involves 9 additions; question 7 involves 5 subtractions, 3 multiplications, and 3 divisions. If out of 20 constituent operations 15 are correct, mark accordingly.

PART I

1. Add 3, 4, 7, 6, 5, 2

2. 20, 35, 47, 84, 73, 10

3. 31,457 − 28,765

4. 72×8

5. 7685×43

6. $6\overline{)4410}$

7. $7\overline{)4,037}$

8. 360, 709, 415, 382

9. $7,460 \times 870$

10. $218\overline{)73,456}$

11. $7046 \times 6\frac{1}{2}$

12. $\frac{1}{2} + \frac{2}{3}$

13. $1\frac{1}{2} - \frac{2}{3} + \frac{1}{4}$

14. $789 - 347\frac{2}{3} + 105\frac{3}{4}$

PART II

1. If a boy buys writing paper for 10¢, pencils for 15¢, an eraser for 5¢, how much change does he receive from 50¢?

2. John sells 35 stamps at 6¢ each. How much does he receive?

3. How many lead pencils at 5¢ each can I buy for 95¢?

4. The product of two numbers is 176; one of the numbers is 16, what is the other?

5. If a man works 8 hours in 1 day, how many hours does he work in 19 days?

6. A farmer sold 75 quarts of cherries at 7¢ a quart. How much money did he receive?

7. A room is 24 ft. long; it is $\frac{3}{4}$ as wide. How wide is it?

8. If 3 tons of coal cost $18, what will $5\frac{1}{2}$ tons cost?

9. John draws a line $5\frac{3}{4}$ inches long and another $3\frac{1}{3}$ inches long. How much longer is the first line than the second?

10. What is the cost of 9 tons of hay at $6.75 a ton?

Lightning Source UK Ltd.
Milton Keynes UK
UKHW011642160119
335572UK00013B/1398/P